INTRODUCTION TO
STATISTICAL PHYSICS

Second Edition

INTRODUCTION TO
STATISTICAL PHYSICS

Second Edition

Kerson Huang

CRC Press
Taylor & Francis Group
Boca Raton   London   New York

CRC Press is an imprint of the
Taylor & Francis Group   an **informa** business

A CHAPMAN & HALL BOOK

Chapman & Hall/CRC
Taylor & Francis Group
6000 Broken Sound Parkway NW, Suite 300
Boca Raton, FL 33487-2742

© 2010 by Taylor and Francis Group, LLC
Chapman & Hall/CRC is an imprint of Taylor & Francis Group, an Informa business

No claim to original U.S. Government works

ISBN 13: 978-1-4200-7902-9 (hbk)

Library of Congress Cataloging-in-Publication Data

Huang, Kerson, 1928-
    Introduction to statistical physics / Kerson Huang. -- 2nd ed.
        p. cm.
    Includes bibliographical references and index.
    ISBN 978-1-4200-7902-9 (hardcover : alk. paper)
    1. Statistical physics. I. Title.

QC174.8.H82 2010
530.15'95--dc22   2009031930

**Visit the Taylor & Francis Web site at
http://www.taylorandfrancis.com**

**and the CRC Press Web site at
http://www.crcpress.com**

To the memory of Herman Feshbach (1917–2000)

Contents

Preface xiii

1 A Macroscopic View of Matter 1
   1.1 Viewing the World at Different Scales 1
   1.2 Thermodynamics 2
   1.3 The Thermodynamic Limit 3
   1.4 Thermodynamic Transformations 4
   1.5 Classic Ideal Gas 7
   1.6 First Law of Thermodynamics 8
   1.7 Magnetic Systems 9
   Problems 11
   References 13

2 Heat and Entropy 15
   2.1 The Heat Equations 15
   2.2 Applications to Ideal Gas 16
   2.3 Carnot Cycle 19
   2.4 Second Law of Thermodynamics 20
   2.5 Absolute Temperature 21
   2.6 Temperature as Integrating Factor 22
   2.7 Entropy 25
   2.8 Entropy of Ideal Gas 26
   2.9 The Limits of Thermodynamics 27
   Problems 27

3 Using Thermodynamics 33
   3.1 The Energy Equation 33
   3.2 Some Measurable Coefficients 34
   3.3 Entropy and Loss 35
   3.4 *TS* Diagram 37
   3.5 Condition for Equilibrium 39
   3.6 Helmholtz Free Energy 40
   3.7 Gibbs Potential 41
   3.8 Maxwell Relations 42
   3.9 Chemical Potential 42
   Problems 43

4   Phase Transitions **47**
    4.1   First-Order Phase Transition 47
    4.2   Condition for Phase Coexistence 49
    4.3   Clapeyron Equation 50
    4.4   Van der Waals Equation of State 51
    4.5   Virial Expansion 53
    4.6   Critical Point 53
    4.7   Maxwell Construction 55
    4.8   Scaling 56
    4.9   Nucleation and Spinodal Decomposition 57
    Problems 60
    References 63

5   The Statistical Approach **65**
    5.1   The Atomic View 65
    5.2   Random Walk 67
    5.3   Phase Space 69
    5.4   Distribution Function 70
    5.5   Ergodic Hypothesis 72
    5.6   Statistical Ensemble 72
    5.7   Microcanonical Ensemble 73
    5.8   Correct Boltzmann Counting 74
    5.9   Distribution Entropy: Boltzmann's H 76
    5.10 The Most Probable Distribution 77
    5.11 Information Theory: Shannon Entropy 78
    Problems 80
    References 82

6   Maxwell–Boltzmann Distribution **83**
    6.1   Determining the Parameters 83
    6.2   Pressure of Ideal Gas 84
    6.3   Equipartition of Energy 85
    6.4   Distribution of Speed 87
    6.5   Entropy 88
    6.6   Derivation of Thermodynamics 89
    6.7   Fluctuations 90
    6.8   The Boltzmann Factor 91
    6.9   Time's Arrow 92
    Problems 93
    References 97

7   Transport Phenomena **99**
    7.1   Collisionless and Hydrodynamic Regimes 99
    7.2   Maxwell's Demon 101
    7.3   Nonviscous Hydrodynamics 101

7.4 Sound Wave 103
7.5 Diffusion 103
7.6 Heat Conduction 105
7.7 Viscosity 106
7.8 Navier–Stokes Equation 107
Problems 109
References 110

8 Canonical Ensemble **111**
8.1 Review of the Microcanonical Ensemble 111
8.2 Classical Canonical Ensemble 111
8.3 The Partition Function 114
8.4 Connection with Thermodynamics 114
8.5 Energy Fluctuations 115
8.6 Minimization of Free Energy 116
8.7 Classical Ideal Gas 118
Problems 119

9 Grand Canonical Ensemble **123**
9.1 The Particle Reservoir 123
9.2 Grand Partition Function 123
9.3 Number Fluctuations 124
9.4 Connection with Thermodynamics 125
9.5 Parametric Equation of State and Virial Expansion 126
9.6 Critical Fluctuations 127
9.7 Pair Creation 128
Problems 130

10 Noise **133**
10.1 Thermal Fluctuations 133
10.2 Nyquist Noise 134
10.3 Brownian Motion 136
10.4 Einstein's Theory 138
10.5 Diffusion 140
10.6 Einstein's Relation 142
10.7 Molecular Reality 143
10.8 Fluctuation and Dissipation 144
10.9 Brownian Motion of the Stock Market 145
Problems 148
References 149

11 Stochastic Processes **151**
11.1 Randomness and Probability 151
11.2 Binomial Distribution 152
11.3 Poisson Distribution 154

11.4   Gaussian Distribution..155
11.5   Central Limit Theorem 157
11.6   Shot Noise 157
Problems 160
References 162

12   Time-Series Analysis **163**
12.1   Ensemble of Paths...163
12.2   Ensemble Average 164
12.3   Power Spectrum and Correlation Function 165
12.4   Signal and Noise..168
12.5   Transition Probabilities 170
12.6   Markov Process 171
12.7   Fokker–Planck Equation 172
12.8   The Monte Carlo Method 173
12.9   Simulation of the Ising Model 176
Problems 179
References 181

13   The Langevin Equation **183**
13.1   The Equation and Solution 183
13.2   Energy Balance 185
13.3   Fluctuation-Dissipation Theorem 187
13.4   Diffusion Coefficient and Einstein's Relation..............187
13.5   Transition Probability: Fokker–Planck Equation 188
13.6   Heating by Stirring: Forced Oscillator in Medium 189
Problems 192

14   Quantum Statistics **195**
14.1   Thermal Wavelength 195
14.2   Identical Particles 197
14.3   Occupation Numbers 198
14.4   Spin 200
14.5   Microcanonical Ensemble..................................201
14.6   Fermi Statistics 202
14.7   Bose Statistics 203
14.8   Determining the Parameters 204
14.9   Pressure 205
14.10 Entropy 206
14.11 Free Energy 207
14.12 Equation of State..207
14.13 Classical Limit...208
Problems 210
Reference 212

15 Quantum Ensembles **213**
   15.1 Incoherent Superposition of States 213
   15.2 Density Matrix 214
   15.3 Canonical Ensemble (Quantum-Mechanical) 216
   15.4 Grand Canonical Ensemble (Quantum-Mechanical) 217
   15.5 Occupation Number Fluctuations 219
   15.6 Photon Bunching 220
   Problems 221
   References 223

16 The Fermi Gas **225**
   16.1 Fermi Energy 225
   16.2 Ground State 226
   16.3 Fermi Temperature 227
   16.4 Low-Temperature Properties 228
   16.5 Particles and Holes 230
   16.6 Electrons in Solids 231
   16.7 Semiconductors 233
   Problems 235

17 The Bose Gas **237**
   17.1 Photons 237
   17.2 Bose Enhancement 239
   17.3 Phonons 241
   17.4 Debye Specific Heat 243
   17.5 Electronic Specific Heat 244
   17.6 Conservation of Particle Number 245
   Problems 246
   References 249

18 Bose–Einstein Condensation **251**
   18.1 Macroscopic Occupation 251
   18.2 The Condensate 253
   18.3 Equation of State 254
   18.4 Specific Heat 256
   18.5 How a Phase is Formed 257
   18.6 Liquid Helium 259
   Problems 260
   References 263

19 The Order Parameter **265**
   19.1 The Essence of Phase Transitions 265
   19.2 Ginsburg–Landau Theory 266
   19.3 Relation to Microscopic Theory 267
   19.4 Functional Integration and Differentiation 268

19.5   Second-Order Phase Transition 270
19.6   Mean-Field Theory 271
19.7   Critical Exponents 273
19.8   The Correlation Length 274
19.9   First-Order Phase Transition 277
19.10 Cahn–Hilliard Equation 278
Problems 278
References 280

20 Superfluidity **281**
20.1   Condensate Wave Function 281
20.2   Spontaneous Symmetry Breaking 282
20.3   Mean-Field Theory 284
20.4   Observation of Bose–Einstein Condensation 285
20.5   Quantum Phase Coherence 286
20.6   Superfluid Flow 287
20.7   Phonons: Goldstone Mode 289
Problems 290
References 292

21 Superconductivity **293**
21.1   Meissner Effect 293
21.2   Magnetic Flux Quantum 294
21.3   Josephson Junction 296
21.4   DC Josephson Effect 298
21.5   AC Josephson Effect 299
21.6   Time-Dependent Vector Potential 300
21.7   The SQUID 300
21.8   Broken Symmetry 302
Problems 303
References 303

Appendix **305**

Index **313**

Preface

The main purpose of statistical physics is to clarify the properties of matter in aggregate, in terms of the physical laws governing atomic motion. The present book is a textbook for advanced undergraduates. It assumes background knowledge of classical and quantum physics, on an introductory undergraduate level.

In introducing the properties of matter, one faces a dilemma: should one follow thermodynamics, and bring in entropy as that mystical object looming out of the second law, or should one first discuss the randomness of atomic collisions? I choose thermodynamics, because it describes everyday experience. (And it is profound, as one realizes after having absorbed the atomic view.)

This book may be divided into three parts, the classical view, the quantum view, and advanced applications.

The classical portion occupies first thirteen chapters, more than half the book. It covers the classical ensembles of statistical mechanics and stochastic processes. The latter subject includes Brownian motion, probability theory, and the Fokker–Planck and Langevin equations, with emphasis on physical understanding. To illustrate the use of statistical methods beyond the theory of matter, there are brief discussions of entropy in information theory, Brownian motion in the stock market, and the Monte Carlo method in computer simulations.

The quantum part comprises five chapters. On quantum ensembles, the discussion emphasizes what makes quantum mechanics different from classical mechanics—the quantum phase. Applications include Fermi statistics and semiconductors, and Bose statistics and Bose–Einstein condensation.

The final three chapters deal with advanced topics. A long chapter introduces what might be viewed as a major phenomenology after thermodynamics—the Ginsburg–Landau theory of the order parameter. The last two chapters are devoted to the special kind of quantum order as manifested in superfluidity and superconductivity.

The present edition is expanded from an earlier edition, which was based on a one-semester course given at MIT. I am grateful for the opportunity to interact with students who took the course, known in MIT lingo as 8.08, and people who helped me teach the course: Alexander Lomakin, Patrick Lee, and Lisa Randall.

Kerson Huang
2009

Chapter 1

A Macroscopic View of Matter

1.1 Viewing the World at Different Scales

The world puts on different faces for observers using different scales in the measurement of space and time. The everyday, macroscopic world, as perceived on the scale of meters and seconds, looks very different from that of the atomic, microscopic world, which is seen on scales smaller by some ten orders of magnitude. Different still is the submicroscopic realm of quarks, which is revealed only when the scale shrinks further by another ten orders of magnitude. With an expanding scale in the opposite direction, one enters the regime of astronomy, and ultimately cosmology. These different pictures arise from different ways of organizing data, while the basic laws of physics remain the same.

The physical laws at the smallest accessible length scale are the most "fundamental," but they are of little use on a larger scale, where we must deal with different physical variables. A complete knowledge of quarks tells us nothing about the structure of nuclei, unless we can define nuclear variables in terms of quarks, and obtain their equations of motion from those for quarks. Needless to say, we are unable to do this in detail, although we can see how this could be done in principle. Therefore, each regime of scales has to be described phenomenologically, in terms of variables and laws observable in that regime. For example, an atom is described in terms of electrons and nuclei without reference to quarks. The subnuclear world enters the equations only implicitly, through such parameters as the mass ratio of electron to proton, which we take from experiments in the atomic regime. Similarly, in the macroscopic domain, where atoms are too small to be visible, we describe matter in terms of phenomenological variables such as pressure and temperature. The atomic structure of matter enters the picture implicitly, in terms of properties such as density and heat capacity, which can be measured by macroscopic instruments.

This book is concerned mainly with statistical methods, which provide a bridge between the microscopic and the macroscopic world. We begin our study with thermodynamics, because it is a highly successful phenomenological theory, which identifies the correct macroscopic variables to use, and serves as a guidepost for statistical mechanics.

1.2 Thermodynamics

From experience, we know that a macroscopic body generally settles down, or "relaxes" to a stationary state after a short time. We call this a state of *thermal equilibrium*. When the external condition is changed, the existing equilibrium state will change, and, after a relatively short relaxation time, settles down to another equilibrium state. Thus, a macroscopic body spends most of the time in some state of equilibrium, punctuated by almost sudden transitions. In our study of macroscopic phenomena, we divide the subject roughly under the following headings:

* *Thermodynamics* is a phenomenological theory of equilibrium states and transitions among them.
* *Statistical mechanics* is concerned with deducing the thermodynamic properties of a macroscopic system from its microscopic structure.
* *Kinetic theory* aims at a microscopic description of the transition process between equilibrium states.

As a rule, properties of a macroscopic system can be classified as either extensive or intensive:

* *Extensive* quantities are proportional to the amount of stuff present.
* *Intensive* quantities are independent of the amount of stuff present.

Generally, there are only these two categories, because we can neglect surface effects. A macroscopic body is typically of size $L \sim 1$ m, while the range of atomic forces is of order $r_0 \sim 10^{-10}$ m. The macroscopic nature is expressed by the ratio $L/r_0 \sim 10^{10}$. The surface to volume ratio, rendered dimensionless in terms of the range of atomic forces, is of order $r_0/L \sim 10^{-10}$. The extensive property expresses the "saturation property" of atomic forces, that is, an atom can "feel" only as far as the range of the force. The intensive property means that atoms in the interior of the body do not feel the presence of the surface.

Exceptions arise when one of the following conditions prevail:

* The system is small.
* There is a nonuniform external potential.
* There are long-range interparticle forces, such as the Coulomb repulsion between charges, and the gravitational attraction between mass elements.
* The geometry is such that the surface is important.

These exceptions occur in important physical systems. For example, the volume of a star is a nonlinear function of its mass, due to the long-ranged gravitational interaction. We illustrate the different cases in Figure 1.1.

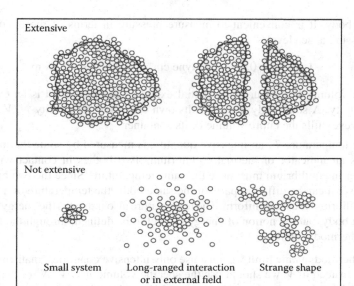

Figure 1.1 A body has extensive properties if surface effects can be neglected, so that the energy is proportional to the number of particles. In these pictures, the surface layer is indicated by a heavy line.

1.3 The Thermodynamic Limit

We consider a material body consisting of N atoms in volume V, in the absence of a nonuniform external potential, to be the idealized limit

$$N \rightarrow \infty$$
$$V \rightarrow \infty$$
$$\frac{N}{V} = \text{fixed number} \tag{1.1}$$

This is called the *thermodynamic limit*, in which the system becomes translationally invariant.

The *thermodynamic state* is specified by a number of thermodynamic variables, which are assumed to be either extensive (proportional to N), or intensive (independent of N). We consider a generic system described by the three variables P, V, T, denoting, pressure, volume, and temperature respectively:

- The pressure P, an intensive quantity, is the force per unit area that the body exerts on a wall, which can be that of the container of the system, or it may be one side of an imaginary surface inside the body. Under equilibrium conditions in the absence of external potentials, the pressure must be uniform throughout the body. It is convenient to measure pressure in terms of the atmospheric pressure at sea level:

$$1 \text{ atm} = 1.013 \times 10^6 \text{ dyne cm}^{-2} = 1.103 \times 10^5 \text{ N m}^{-2} \qquad (1.2)$$

- The volume V measures the spatial extent of the body, and is an extensive quantity. A solid body maintains its own characteristic density N/V. A gas, however, fills the entire volume of its container.
- The temperature T, an intensive quantity, is measured by some thermometer. It is an indicator of thermal equilibrium. Two bodies in contact with each other in equilibrium must have the same temperature. Since the two bodies in question can be different parts of the same body, the temperature of a body in equilibrium must be uniform. The temperature also indicates the energy content of a body, but the notion of energy has yet to be defined, through the first law of thermodynamics.

In the thermodynamic limit we must use only intensive quantities, mathematically speaking. Instead of V we should use the specific volume $v = V/N$, or the density $n = N/V$. However, it is convenient to regard V as a large but finite number, for this corresponds to the everyday experience of seeing the volume of a macroscopic body expand or contract, while the number of atoms is fixed.

There are systems requiring other variables in addition to, or in place of, P, V, T. Common examples are the magnetic field and magnetization for a magnetic substance, the strain and stress in elastic solids, or the surface area and the surface tension.

1.4 Thermodynamic Transformations

When a body is in thermal equilibrium, the thermodynamic variables are not independent of one another, but are constrained by an *equation of state* of the form

$$f(P, V, T) = 0 \qquad (1.3)$$

where the function f is characteristic of the substance. This leaves two independent variables out of the original three. Geometrically we can represent the equation of state by a surface in the state space spanned by P, V, T, as shown in Figure 1.2. All equilibrium states must lie on this surface. We regard f as a continuous differentiable function, except possibly at special points.

A change in the external condition will change the equilibrium state of a system. For example, application of external pressure will cause the volume of a body to decrease. Such a change is called a *thermodynamic transformation*. The initial and final states are equilibrium states. The system can be considered to remain in equilibrium, if the transformation proceeds sufficiently slowly. In such a case, we say that the transformation is *quasi-static*. This usually means that the transformation is *reversible*, in that the system will retrace the transformation in reverse, when the external change

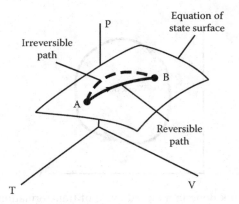

Figure 1.2 The state space in thermodynamics.

is reversed. A reversible transformation can be represented by a continuous path on the equation-of-state surface, as illustrated in Figure 1.2.

An *irreversible* transformation, on the other hand, cannot be represented by a path on the equation-of-state space, as indicated by the dashed line in Figure 1.2. In fact, we may not be able to represent it as a path in the state space at all. An example is the sudden removal of a wall in a container of a gas, so that the gas expands into a compartment that was originally vacuous. Although the initial and final states are equilibrium states, the intermediate states do not have uniform P, V, T, and hence cannot be represented as points in the state space.

In a reversible transformation, we can consider mathematically infinitesimal paths. The mechanical work done by the system over an infinitesimal path is represented by a differential:

$$dW = PdV \tag{1.4}$$

Along a finite reversible path $A \rightarrow B$, the work done is given by

$$\Delta W = \int_A^B PdV \tag{1.5}$$

which depends on the path connecting A to B. This is the area underneath the path in a PV diagram. When the path is a closed cycle, the work done in one cycle is the area enclosed, as shown in Figure 1.3. The work done along an irreversible path is generally not $\int PdV$. For example, in the free expansion of a gas into a vacuum, the system does not perform work on any external agent, and so $\Delta W = 0$.

A uniquely thermodynamic process is heat transfer. From the atomic point of view, it represents a transfer of energy in the form of thermal agitation. In thermodynamics, we define heat phenomenologically, as that imparted by a heating element, such as a flame or a heating coil. An amount of heat ΔQ absorbed by a body causes a rise ΔT in its temperature given by

$$\Delta Q = C \, \Delta T \tag{1.6}$$

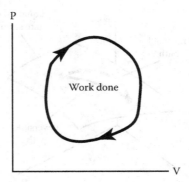

Figure 1.3 The work done in a closed cycle of transformations is represented by the area enclosed by the cycle in a *PV* diagram.

where C is the *heat capacity* of the substance. We imagine the limit in which ΔQ and ΔT become infinitesimal. The heat capacity is an extensive quantity. The intensive heat capacity per particle C/N, per mole C/n, or per unit volume C/V, is called *specific heat.*

The fact that heat is a form of energy was established experimentally, by observing that one can increase the temperature of a body by ΔT either by transferring heat to the body or performing work on it. A practical unit for heat is the *calorie (cal)*, originally defined as that amount of heat that will raise the temperature of 1 g of water from 14.5 to 15.5°C at sea level. In current usage, it is defined exactly in terms of the *joule* (J):

$$1 \text{ cal} \equiv 4.184 \text{ J} \tag{1.7}$$

Another commonly used unit is the *British thermal unit* (Btu):

$$1 \text{ Btu} \equiv 1055 \text{ J} \tag{1.8}$$

The heat absorbed by a body depends on the path of the transformation, as is true of the mechanical work done by the body. We can speak of the amount of heat absorbed in a process, but the "heat of a body," like the "work of a body," is meaningless. Commonly encountered transformations are the following:

- $T = $ constant (isothermal process)
- $P = $ constant (isobaric process)
- $V = $ constant (constant-volume process)
- $\Delta Q = 0$ (adiabatic process)

We use a subscript to distinguish the different types of paths, as for example C_V and C_P, representing, respectively, the heat capacity at constant volume and constant pressure. The heat capacity is only one of many thermodynamic coefficients that measure the response of the system to an external source. Other examples are

$$\kappa = -\frac{1}{V}\frac{\Delta V}{\Delta P} \quad \text{(compressibility)}$$

$$\alpha = \frac{1}{V}\frac{\Delta V}{\Delta T} \quad \text{(coefficient of thermal expansion)} \quad (1.9)$$

These coefficients can be obtained from experimental measurements. In principle they can be calculated from atomic properties using statistical mechanics.

1.5 Classic Ideal Gas

The simplest thermodynamic system is the classic ideal gas, which is a gas in the limit of low density and high temperature. The equation of state is given by the *ideal gas law*:

$$PV = Nk_BT \quad (1.10)$$

where T is the *ideal gas temperature*, measured in kelvins (K), and

$$k_B = 1.381 \times 10^{-16} \text{ erg K}^{-1} \text{ (Boltzmann's constant)} \quad (1.11)$$

As we shall see, the second law of thermodynamics implies $T > 0$, and the lower bound is called the *absolute zero*. For this reason, T is also called the *absolute temperature*. The heat capacity of a monatomic ideal gas at constant volume C_V has the value

$$C_V = \frac{3}{2}Nk_B \quad (1.12)$$

These properties of the ideal gas were established experimentally, and can be derived theoretically in statistical mechanics.

Thermodynamics does not assume the existence of atoms. Instead of the number of atoms N, we can use the number of gram moles n, which is a chemical property of the substance. The two are related through

$$Nk_B = nR$$

$$R = 8.314 \times 10^7 \text{ erg K}^{-1} \text{ (gas constant)} \quad (1.13)$$

The ratio R/k_B is Avogadro's number, the number of atoms per mole:

$$A_0 = \frac{R}{k_B} = 6.022 \times 10^{23} \quad (1.14)$$

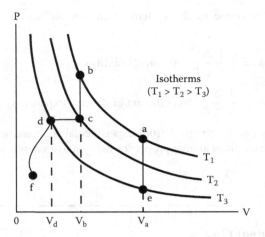

Figure 1.4 Isotherms of an ideal gas, and various paths of transformations.

Indeed, the atomic picture was a latecomer that gained acceptance only after a long historic struggle (See Section 10.7).

The equation of state can be represented graphically in a *PV* diagram, as shown in Figure 1.4, which displays a family of curves at constant *T* called *isotherms*. Indicated on this graph are the reversible paths corresponding to various types of transformations:

- *ab* is isothermal.
- *bc* proceeds at constant volume.
- *cd* is at constant pressure.
- *de* is isothermal.
- *abcdea* is a closed cycle.
- *df* is nonisothermal.

To keep the temperature constant during an isothermal transformation, we keep the system in thermal contact with a body so large that its temperature is not noticeably affected by heat exchange with our system. We call such a body a *heat reservoir*, or *heat bath*.

1.6 First Law of Thermodynamics

The first law expresses the conservation of energy by including heat as a form of energy. It asserts that there exists a function of the state, *internal energy U*, whose change in any thermodynamic transformation is given by

$$\Delta U = \Delta Q - \Delta W \tag{1.15}$$

That is, ΔU is independent of the path of the transformation, although ΔQ and ΔW are path dependent. In a reversible infinitesimal transformation, the infinitesimal changes dQ and dW are not exact differentials, in the sense that they do not represent the changes of definite functions, but their difference

$$dU = dQ - dW \qquad (1.16)$$

is an exact differential.

1.7 Magnetic Systems

For a magnetic substance, the thermodynamic variables are the intensive magnetic field H, and the extensive magnetization M, which are unidirectional and uniform in space. The magnetic field is generated by external real currents (and not induced currents), and the magnet work done by the system is given by

$$dW = -HdM \qquad (1.17)$$

The first law takes the form

$$dU = dQ - HdM \qquad (1.18)$$

which maps into that for a PVT system under the correspondence $H \leftrightarrow -P, M \leftrightarrow V$.

For a paramagnetic substance, a magnetic field induces a magnetization density given by

$$\frac{M}{V} = \chi H \qquad (1.19)$$

The magnetic susceptibility χ obeys Curie's law

$$\chi = \frac{c_0}{T} \qquad (1.20)$$

Thus, we have the equation of state

$$M = \frac{\kappa H}{T} \qquad (1.21)$$

where $\kappa = c_0 V$.

An idealized uniform ferromagnetic system has a phase transition at a critical temperature T_c, and becomes a permanent magnet for $T < T_c$ in the absence of external field. However, there is no phase transition in the presence of an external field. The equation of state is illustrated graphically in the MT diagram of Figure 1.5.

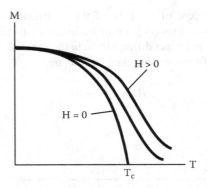

Figure 1.5   Equation of state of a uniform ferromagnet. A ferromagnetic transition occurs at a critical temperature at zero field.

A real ferromagnet is not uniform, but made up of domains with different orientations of the magnetization. The configuration of the domains is such as to reduce the fringing fields that extend outside of the body, in order to minimize the magnetic energy. Each domain behaves according to the uniform case depicted in Figure 1.5. At a fixed $T < T_c$, the domain walls move in response to a change in H, to either consolidate old domains, or create new ones. This is a dissipative process, and leads to hysteresis, as shown in the *MH* plot of Figure 1.6. At zero field, the magnetization is at either a or b, depending on the history. This is the basis of magnetic memory.

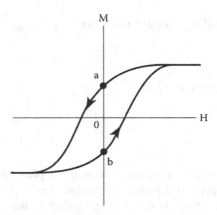

Figure 1.6   Hysteresis in a ferromagnet is due to the formation of ferromagnetic domains.

Problems

1.1 Tenzing Norgay heated 1 m³ of water from 20 to 40°C, in order to take a bath before climbing Chomolungma.

(a) How much energy did he use? Give the answer in kWh.

(b) Show that the energy is enough to lift Tenzing's entire party of 14, of average weight 150 lb, from sea level to the top of Chomolungma (elev. 29,000 ft).

1.2 Referring to Figure 1.4, find the work done along the various paths on the closed cycle ab, bc, cd, de, ea, and give the total work done in the closed cycle. How much heat is supplied to the system in one cycle?

1.3 An ideal gas undergoes a reversible transformation along the path

$$\frac{V}{V_0} = \left(\frac{T}{T_0}\right)^b$$

where V_0, T_0, and b are constants.

(a) Find the coefficient of thermal expansion.

(b) Calculate the work done by the gas when the temperature increases by ΔT.

1.4 In a nonuniform gas, the equation of state is valid locally, as long as the density does not change too rapidly with position. Consider a column of ideal gas under gravity at constant temperature. Find the density as a function of height, by balancing the forces acting on a volume element.

1.5 Two solid bodies labeled 1 and 2 are in thermal contact with each other. The initial temperatures were T_1, T_2, with $T_1 > T_2$. The heat capacities are C_1 and C_2, respectively. What is is final equilibrium temperature, if the bodies are completely isolated from the external world?

1.6 A paramagnetic substance is magnetized in an external magnetic field at constant temperature. How much work is require to attain a magnetization M?

1.7 A hysteresis curve (see Figure 1.6) is given by the formula

$$M = M_0 \tanh(H \pm H_0)$$

where the $+$ sign refers to the upper branch, and the $-$ sign refers to the lower branch. The parameter H_0 is called the *coercive force*. Show that the work done by the system in one hysteresis cycle is $2M_0 H_0$.

1.8 The atomic nucleus The atomic nucleus contains typically fewer than 300 protons and neutrons. It is an example of a small system with both short- and long-range forces. The mass is given by the semi empirical Weizsäcker formula (See deShalit and Feshbach 1974):

$$M = Zm_p + Nm_n - a_1 A + a_2 A^{2/3} + a_3 \frac{Z^2}{A^{1/3}} + a_4 \frac{(Z-N)^2}{A} + \delta(A)$$

where Z is the number of protons, N is the number of neutrons, and $A = Z + N$. The masses of protons and neutrons are, respectively, m_p, m_n, and a_i $(i = 1, \ldots, 4)$ are numerical constants.

Volume energy : $a_1 = 16$ MeV/c^2

Surface energy : $a_2 = 19$

Coulomb energy : $a_3 = 0.72$

Symmetry energy : $a_4 = 28$

The symmetry energy favors $N = Z$. The term $\delta(A)$ is the "pairing energy" that gives small fluctuations.

Take $Z = N = A/2$, $m_n \approx m_p$, and neglect $\delta(A)$, so that

$$M = (m_p - a_1)A + a_2 A^{2/3} + \frac{a_3}{4} A^{5/3}$$

Make a log plot of M from $A = 1$ to $A = 10^4$, and indicate the range in which M can be considered as extensive.

1.9 The false vacuum In the theory of the "inflationary universe," a pinpoint universe was created during the Big Bang in a "false" vacuum state characterized by a finite energy density u_0, while the true vacuum state should have zero energy density. (For a review see Guth, 1992).

An ordinary solid too has finite energy per unit volume; but it occupies a definite volume, given a total number of particles. If we put the solid in a box larger than that volume, it would not fill the box, but "rattle" in it. Not so the false vacuum. It must have the same energy density whatever its volume, and this makes it strange stuff indeed.

(a) Imagine putting the false vacuum in a cylinder stopped with a piston, with the true vacuum outside. If we pull on the piston to increase the volume of the false vacuum by dV, show that the false vacuum performs work $dW = -u_0 dV$. Hence the false vacuum has negative pressure $P = -u_0$.

(b) According to general relativity, the radius R of the universe obeys the equation

$$\frac{d^2 R}{dt^2} = -\frac{4\pi}{3c^2} G(u_0 - 3P)R$$

where c is the velocity of light, and

$$G = 6.673 \times 10^{-8} \text{cm}^3/\text{g s} = 6.707 \times 10^{-39} \text{ GeV}^{-2}\hbar c^3$$

is Newton's constant (gravitational constant). For the false vacuum, this reduces to

$$\frac{d^2 R}{dt^2} = \frac{R}{\tau^2}$$

$$\tau = \sqrt{\frac{3c^2}{8\pi G u_0}}$$

Thus the radius of the universe expands exponentially with time constant τ. According to theory, $u_0 = (10^{15}\text{GeV})^4/(\hbar c)^3 = 10^{96}$ erg/cm^3. Show

$$\tau \approx 10^{-34}s$$

References

deShalit, A. and H. Feshbach, *Theoretical Nuclear Physics*, Vol. 1, Wiley, New York, 1974, p. 11.

Guth, A., *The Oskar Klein Memorial Lectures*, Vol. 2, G. Ekspong (ed.), World Scientific Publishing Co., Singapore, 1992.

Chapter 2

Heat and Entropy

2.1 The Heat Equations

Thermodynamics becomes a mathematical science when we regard the state functions, such as the internal energy, as continuous differentiable functions of the variables P, V, T. The constraint imposed by the equation of state reduces the number of independent variables to two. We may consider the internal energy to be a function of any two of the variables. Under infinitesimal increments of the variables, we can write

$$dU(P, V) = \left(\frac{\partial U}{\partial P}\right)_V dP + \left(\frac{\partial U}{\partial V}\right)_P dV$$

$$dU(P, T) = \left(\frac{\partial U}{\partial P}\right)_T dP + \left(\frac{\partial U}{\partial T}\right)_P dT$$

$$dU(V, T) = \left(\frac{\partial U}{\partial V}\right)_T dV + \left(\frac{\partial U}{\partial T}\right)_V dT \tag{2.1}$$

where a subscript on a partial derivative denotes the variable being held fixed. For example, $(\partial U/\partial P)_V$ is the derivative with respect to P at constant V. These partial derivatives are thermodynamic coefficients to be taken from experiments.

The heat absorbed by the system can be obtained from the first law, written in the form $dQ = dU + PdV$:

$$dQ = \left(\frac{\partial U}{\partial P}\right)_V dP + \left[\left(\frac{\partial U}{\partial V}\right)_P + P\right] dV$$

$$dQ = \left(\frac{\partial U}{\partial P}\right)_T dP + \left(\frac{\partial U}{\partial T}\right)_P dT + PdV$$

$$dQ = \left[\left(\frac{\partial U}{\partial V}\right)_T + P\right] dV + \left(\frac{\partial U}{\partial T}\right)_V dT \tag{2.2}$$

In the second of these equations we must regard V a function of P, T, and rewrite

$$dV = \left(\frac{\partial V}{\partial P}\right)_T dP + \left(\frac{\partial V}{\partial T}\right)_P dT \tag{2.3}$$

15

The second equation then reads

$$dQ = \left[\left(\frac{\partial U}{\partial P} \right)_T + P \left(\frac{\partial V}{\partial P} \right)_T \right] dP + \left(\frac{\partial (U + PV)}{\partial T} \right)_P dT \qquad (2.4)$$

It is convenient to define a state function called the *enthalpy*:

$$H \equiv U + PV \qquad (2.5)$$

The heat equations in dQ form are summarized below:

$$dQ = \left(\frac{\partial U}{\partial P} \right)_V dP + \left[\left(\frac{\partial U}{\partial V} \right)_P + P \right] dV$$

$$dQ = \left[\left(\frac{\partial U}{\partial P} \right)_T + P \left(\frac{\partial V}{\partial P} \right)_T \right] dP + \left(\frac{\partial H}{\partial T} \right)_P dT$$

$$dQ = \left[\left(\frac{\partial U}{\partial V} \right)_T + P \right] dV + \left(\frac{\partial U}{\partial T} \right)_V dT \qquad (2.6)$$

We can immediately read off the heat capacities at constant V and P:

$$C_V = \left(\frac{\partial U}{\partial T} \right)_V$$

$$C_P = \left(\frac{\partial H}{\partial T} \right)_P \qquad (2.7)$$

These are useful because they express the heat capacities as derivatives of state functions.

2.2 Applications to Ideal Gas

We now use the first law to deduce some properties of the ideal gas. Joule performed a classic experiment on free expansion, as illustrated in Figure 2.1. A thermally insulated ideal gas was allowed to expand freely into an insulated chamber, which was initially vacuous. After a new equilibrium was established, in which the gas fills both compartments, the final temperature was found to be the same as the initial temperature.

We can draw the following conclusions:

- $\Delta W = 0$ (since the gas pushes into a vacuum)
- $\Delta Q = 0$ (since the temperature was unchanged)
- $\Delta U = 0$ (by the first law)

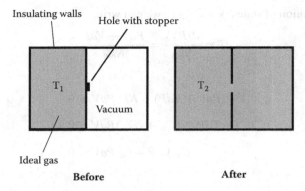

Figure 2.1   Free expansion of an ideal gas.

Choosing V, T as independent variables, we conclude $U(V_1, T) = U(V_2, T)$. That is, U is independent of V:

$$U = U(T) \qquad (2.8)$$

Of course, U is proportional to the total number of particle N, which has been kept constant.

The heat capacity at constant volume can now be written as a total derivative:

$$C_V = \left(\frac{\partial U}{\partial T} \right)_V = \frac{dU}{dT} \qquad (2.9)$$

Assuming that C_V is a constant, we can integrate the above to obtain

$$U(T) = \int C_V \, dT = C_V T \qquad (2.10)$$

where the constant of integration has been set to zero by defining $U = 0$ at $T = 0$. It follows that

$$C_P = \left(\frac{\partial H}{\partial T} \right)_P = \left(\frac{\partial (U + PV)}{\partial T} \right)_P$$

$$= \frac{dU}{dT} + \frac{\partial (Nk_B T)}{\partial T} \qquad (2.11)$$

$$= C_V + Nk_B \qquad (2.12)$$

Thus, for an ideal gas,

$$C_P - C_V = Nk_B \qquad (2.13)$$

We now work out the equation governing a reversible adiabatic transformation. Setting $dQ = 0$, we have $dU = -PdV$. Since $dU = C_V dT$, we obtain

$$C_V dT + PdV = 0 \qquad (2.14)$$

Using the equation of state $PV = Nk_BT$, we can write

$$dT = \frac{d(PV)}{Nk_B} = \frac{PdV + VdP}{Nk_B} \tag{2.15}$$

Thus

$$C_V (PdV + VdP) + Nk_B PdV = 0$$

$$C_V VdP + (C_V + Nk_B)PdV = 0$$

$$C_V VdP + C_P PdV = 0 \tag{2.16}$$

or

$$\frac{dP}{P} + \gamma \frac{dV}{V} = 0 \tag{2.17}$$

where

$$\gamma \equiv \frac{C_P}{C_V} \tag{2.18}$$

Assuming that γ is a constant, we obtain through an integration

$$\ln P = -\gamma \ln V + \text{constant} \tag{2.19}$$

or

$$PV^\gamma = \text{constant} \tag{2.20}$$

Using the equation of state, we can rewrite this in the equivalent form

$$TV^{\gamma-1} = \text{constant} \tag{2.21}$$

Since $\gamma > 1$ according to Equation (2.13), an adiabatic path has a steeper slope than an isotherm in a PV diagram, as depicted in Figure 2.2.

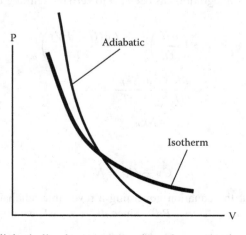

Figure 2.2 An adiabatic line has a steeper slope than an isotherm.

2.3   Carnot Cycle

In a cyclic transformation, the final state is the same as the initial state, and therefore $\Delta U = 0$, because U is a state function. A reversible cyclic process can be represented by a closed loop in the PV diagram. The area of the loop is the total work done by the system in one cycle. Since $\Delta U = 0$, it is also equal to the heat absorbed:

$$\Delta W = \Delta Q = \oint PdV = \text{area enclosed} \tag{2.22}$$

A cyclic process converts work into heat and returns the system to its original state. It acts as a heat engine, for the process can be repeated indefinitely. If the cycle is reversible, it runs as a refrigerator in reverse.

A Carnot cycle is a reversible cycle bounded by two isotherms and two adiabatic lines. The working substance is arbitrary, but we illustrate it for an ideal gas in Figure 2.3, where $T_2 > T_1$. The system absorbs heat Q_2 along the isotherm T_2, and rejects heat Q_1 along T_1, with $Q_1 > 0$ and $Q_2 > 0$. By the first law, the net work output is

$$W = Q_2 - Q_1 \tag{2.23}$$

In one cycle of operation, the system receives an amount of heat Q_2 from a hot reservoir, performs work W, and rejects "waste heat" Q_1 to a cold reservoir. The efficiency of the Carnot engine is defined as

$$\eta \equiv \frac{W}{Q_2} = 1 - \frac{Q_1}{Q_2} \tag{2.24}$$

which is 100% if there is no waste heat, that is, $Q_1 = 0$. But, as we shall see, the second law of thermodynamics states that this is impossible.

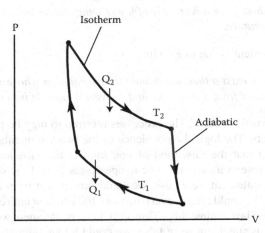

Figure 2.3   Carnot cycle on the PV diagram of an ideal gas.

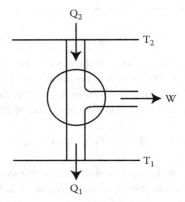

Figure 2.4   Schematic representation of a Carnot engine.

We represent the Carnot engine by the schematic diagram shown in Figure 2.4, which emphasizes the fact that the working substance is irrelevant. By reversing the signs of Q_1 and Q_2, and thus W, we can run the engine in reverse as a Carnot refrigerator.

2.4   Second Law of Thermodynamics

The second law of thermodynamics expresses the common wisdom that "heat does not flow uphill." It is stated more precisely by Clausius:

> *There does not exist a thermodynamic transformation whose sole effect is to deliver heat from a reservoir of lower temperature to a reservoir of higher temperature.*

An equivalent statement is due to Kelvin:

> *There does not exist a thermodynamic transformation whose sole effect is to extract heat from a reservoir and convert it entirely into work.*

The important word is "sole." The processes referred to may be possible, but not without other effects. The logical equivalence of the two statements can be demonstrated by showing that the falsehood of one implies the falsehood of the other. Consider two heat reservoirs at respective temperatures T_2 and T_1, with $T_2 > T_1$.

(a) If the Kelvin statement were false, we could extract heat from T_1 and convert it entirely into work. We could then convert the work back to heat entirely, and deliver it to T_2, (there being no law against this). Thus the Clausius statement would be negated.

(b) If the Clausius statement were false, we could let an amount of heat Q_2 flow uphill, from T_1 to T_2. We could then connect a Carnot engine between T_2 and T_1, to extract Q_2 from T_2, and return an amount $Q_1 < Q_2$ back to T_1. The net work output is $Q_2 - Q_1 > 0$. Thus, an amount of heat $Q_2 - Q_1$ is converted into work entirely, without any other effect. This would contradict the Kelvin statement.

In the atomic view, heat transfer represents an exchange of energy residing in the random motion of the atoms. In contrast, the performance of work requires an organized actions of the atoms. To convert heat entirely work would mean that chaos spontaneously reverts to order. This is extremely improbable, for in the usual scheme of things, only one configuration corresponds to order, while all others lead to chaos. The second law is the thermodynamic way of expressing this idea.

2.5 Absolute Temperature

The second law immediately implies that a Carnot engine cannot be 100% efficient, for otherwise all the heat absorbed from the upper reservoir would be converted into work in one cycle of operation. There is no other effect, since the system returns to its original state.

We can show that no engines working between two given temperatures can be more efficient than a Carnot engine. Since only two reservoirs are present, a Carnot engine simply means a reversible engine. What we assert then, is that an irreversible engine cannot be more efficient than a reversible one.

Consider a Carnot engine C and an engine X (not necessary reversible) working between the reservoirs T_2 and T_1, with $T_2 > T_1$, as shown in Figure 2.5. We shall run C in reverse, as a refrigerator $\bar{C}$, and feed the work output of X to $\bar{C}$. Table 2.1 shows a balance sheet of heat transfer in one cycle of joint operation.

The total work output is

$$W_{\text{tot}} = (Q_2' - Q_1') - (Q_2 - Q_1) \tag{2.25}$$

Now arrange to have $Q_2' = Q_2$. Then, no net heat was extracted from the reservoir T_2, which can be ignored. An amount of heat $Q_1 - Q_1'$ was extracted from the reservoir T_1 and converted entirely into work with no other effect. This would violate the second

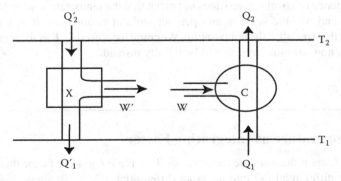

Figure 2.5 Driving a Carnot referigarator $\bar{C}$ with an arbitrary engine X.

TABLE 2.1   Balance Sheet of Heat Transfer

Engine	From T_2	To T_1
X	Q_2'	Q_1'
$\bar{C}$	$-Q_2$	$-Q_1$

law, unless $Q_1 \leq Q_1'$. Dividing both sides of this inequality by Q_2, and using the fact $Q_2' = Q_2$, we have

$$\frac{Q_1}{Q_2} \leq \frac{Q_1'}{Q_2'} \qquad (2.26)$$

Therefore $1 - (Q_1/Q_2) \geq 1 - (Q_1'/Q_2')$, or

$$\eta_C \geq \eta_X \qquad (2.27)$$

As a corollary, all Carnot engines have the same efficiency, since X may be a Carnot engine. This shows that the Carnot engine is universal, in that it depends only on the temperatures involved, and not on the working substance.

We define the *absolute temperature* θ of a heat reservoir such that the ratio of the absolute temperatures of two reservoirs is given by

$$\frac{\theta_1}{\theta_2} \equiv 1 - \frac{Q_1}{Q_2} = 1 - \eta \qquad (2.28)$$

where η is the efficiency of a Carnot engine operating between the two reservoirs. The advantage of this definition is that it is independent of the properties of any working substance. Since $Q_1 > 0$ according to the second law, the absolute temperature is bounded from below:

$$\theta > 0 \qquad (2.29)$$

The *absolute zero* $\theta = 0$ is a limiting value which we can never reach, according to the second law.

The absolute temperature coincides with the ideal gas temperature defined by $T = PV/Nk_B$, as we can show by using an ideal gas as working substance in a Carnot engine. Thus $\theta = T$, and we shall henceforth denote the absolute temperature by T.

The existence of absolute zero does not mean that the temperature scale terminates at the low end, for the scale is an open set without boundaries. It is a matter of convention that we call T the temperature. We could have used $1/T$ as the temperature, and what is now absolute zero would be infinity instead.

2.6   Temperature as Integrating Factor

We can look upon the absolute temperature T as the integrating factor that converts the inexact differential dQ into an exact differential dQ/T. To show this, we first prove a theorem due to Clausius:

In an arbitrary cyclic process P, the following inequality holds:

$$\oint_P \frac{dQ}{T} \leq 0 \qquad (2.30)$$

where the equality holds if P is reversible.

It should be emphasized that the cycle process need not be reversible. To prove the assertion, divide the cycle P into K segments labeled $i = 1, \ldots, K$. Let the ith segment be in contact with a reservoir of temperature T_i, from which it absorbs an amount of heat Q_i. The total work output of P is, by the first law,

$$W = \sum_{i=1}^{K} Q_i \qquad (2.31)$$

Note that not all the Q_i can be positive, for otherwise heat would have been converted to work with no other effect, in contradiction to the second law.

Imagine a reservoir at a temperature $T_0 > T_i$ (all i), with Carnot engines C_i operating between T_0 and each of the temperatures T_i. The setup is illustrated schematically in Figure 2.6. Suppose that, in one cycle of operation, the Carnot engine C_i absorbs heat $Q_i^{(0)}$ from the T_0 reservoir, and rejects Q_i to the T_i reservoir. By definition of the

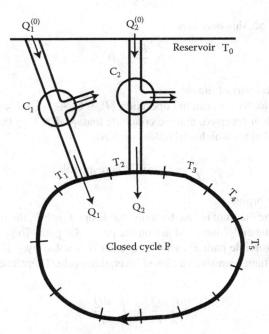

Figure 2.6   Construction to prove Clausius' theorem.

absolute temperature, we have

$$\frac{Q_i^{(0)}}{Q_i} = \frac{T_0}{T_i} \tag{2.32}$$

In one cycle of operation of the joint operations $\{P + C_1 + \cdots + C_K\}$,

- The reservoir T_i experiences no net change, for it receives Q_i, and delivers same to the system.

- The heat extracted from the T_0 reservoir is given by

$$Q_{\text{tot}} = \sum_{i=1}^{K} Q_i^{(0)} = T_0 \sum_{i=1}^{K} \frac{Q_i}{T_i} \tag{2.33}$$

- The total work output is

$$W_{\text{tot}} = W + \sum_{i=1}^{K} \left[Q_i^{(0)} - Q_i \right] = \sum_{i=1}^{K} Q_i^{(0)} = Q_{\text{tot}} \tag{2.34}$$

An amount of heat Q_{tot} would be entirely converted to work with no other effect, and thus violate the second law, unless $Q_{\text{tot}} \leq 0$, or

$$\sum_{i=1}^{K} \frac{Q_i}{T_i} \leq 0 \tag{2.35}$$

In the limit $K \to \infty$, this becomes

$$\oint_P \frac{dQ}{T} \leq 0 \tag{2.36}$$

This proves the first part of the theorem.

If P is reversible, we can run the operation $\{P + C_1 + \cdots + C_N\}$ in reverse. The signs for Q_i are then reversed, and we conclude that $\oint_P dQ/T \geq 0$. Combining this with the earlier relation, which still holds, we have

$$\oint_P \frac{dQ}{T} = 0 \quad \text{(if P is reversible)} \tag{2.37}$$

This complete the proof.

A corollary to the theorem is that, for a reversible open path P, the integral $\int_P dQ/T$ depends only on the endpoints, and not on the particular path. To prove this, join the endpoints by a reversible path P', whose reversal is denoted by $-P'$. The combined processes $P - P'$ then represents a closed reversible cycle. Therefore $\int_{P-P'} dQ/T = 0$, or

$$\int_P \frac{dQ}{T} = \int_{P'} \frac{dQ}{T} \tag{2.38}$$

This shows that dQ divided by the absolute temperature is an exact differential.

2.7 Entropy

The exact differential

$$dS = \frac{dQ}{T} \tag{2.39}$$

defines a state function S called the *entropy*, up to an additive constant. The entropy difference between any two states B and A is given by

$$S(B) - S(A) \equiv \int_A^B \frac{dQ}{T} \text{ (along any reversible path)} \tag{2.40}$$

where the integral extends along any *reversible* path connecting B to A. The result is of course independent of the path, as long such a reversible path exists.

What if we integrate along an irreversible path? Let P be an arbitrary path from A to B, reversible or not. Let R be a reversible path with the same endpoints. The the combined process $P - R$ is a closed cycle, and therefore by Clausius' theorem $\int_{P-R} dQ/T \le 0$, or

$$\int_P \frac{dQ}{T} \le \int_R \frac{dQ}{T} \tag{2.41}$$

Since the right side is the definition of the entropy difference between the final state B and the initial state A, we have

$$S(B) - S(A) \ge \int_A^B \frac{dQ}{T} \tag{2.42}$$

where the equality holds if the process is reversible. For an isolated system, which does not exchange heat with the external world, we have $dQ = 0$, and therefore

$$\Delta S \ge 0 \tag{2.43}$$

That is, the entropy of an isolated system never decreases and it remains constant during a reversible transformation.

We emphasize the following points:

- The principle that the entropy never decreases applies to the "universe" consisting of a system and its environments. It does not apply to a nonisolated system, whose entropy may increase or decrease.

- Since the entropy is a state function, the entropy change of the system in going from state A to state B is $S_B - S_A$ regardless of the path, which may be reversible or irreversible. For an irreversible path, the entropy of the environment changes, whereas for a reversible path it does not change.

- The entropy difference $S_B - S_A$ is not necessarily equal to the integral $\int_A^B dQ/T$. It is equal to the integral only if the path from A to B is reversible. Otherwise, it is generally *larger* than the integral.

2.8 Entropy of Ideal Gas

We can calculate the entropy of an ideal gas as a function of V and T by integrating $dS = dQ/T$. In Figure 2.7, we approach point A along two alternative paths, with V kept fixed along path 1, and T kept fixed along path 2. Along path 1, we have $\int dQ/T = C_V \int dT/T$, and hence

$$S(V, T) = S(V, T_0) + C_V \int_{T_0}^{T} \frac{dT}{T} = S(V, T_0) + C_V \ln \frac{T}{T_0} \qquad (2.44)$$

To determine $S(V, T_0)$, we integrate $dS = dQ/T$ along the isothermal path 2. Since $dU = 0$, we have

$$dQ = dW = PdV = Nk_B T \frac{dV}{V} \qquad (2.45)$$

Thus

$$S(V, T) = S(V_0, T) + Nk_B \int_{V_0}^{V} \frac{dV}{V} = S(V_0, T) + Nk_B \ln \frac{V}{V_0} \qquad (2.46)$$

Comparing the two expressions for $S(V, T)$, we conclude

$$S(V, T_0) = C_0 + Nk_B \ln V \qquad (2.47)$$

where C_0 is an arbitrary constant. Absorbing the constant V_0 into C_0, we can write

$$S(V, T) = C_0 + Nk_B \ln V + C_V \ln T \qquad (2.48)$$

For a monatomic gas we have $C_V = \frac{3}{2}Nk$, and hence

$$S(V, T) = C_0 + Nk_B \ln(VT^{3/2}) \qquad (2.49)$$

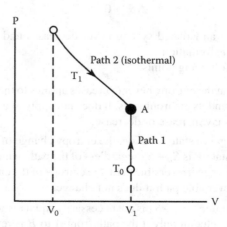

Figure 2.7   Calculating the entropy at point A.

A problem arises when N changes, for S behaves like $N \ln V$, and is not extensive, unless somehow C_0 contains a term $-Nk_B \ln N$. As we shall see, this is supplied by quantum effects, through "correct Boltzmann counting," as explained in Chapters 5 and 8.

Looking forward, we quote the *Sacker–Tetrode equation* for the absolute entropy:

$$S = Nk_B \left[\frac{5}{2} - \ln(n\lambda^3) \right] \qquad (2.50)$$

which will be derived in Chapter 8. Here, $n = N/V$ and λ is the *thermal wavelength*, the deBroglie wavelength of a particle with energy $k_B T$:

$$\lambda = \sqrt{2\pi\hbar^2/mk_B T} \qquad (2.51)$$

where $\hbar$ is the reduced Planck's constant. It is interestig to note that this quantum constant appears even in high-temperature macroscopic physics.

2.9 The Limits of Thermodynamics

Thermodynamics is a very useful practical tool, and an elegant theory. The self-consistency of the mathematical structure has been demonstrated through axiomatic formulations. However, confrontation with experiments indicates that thermodynamics is valid only insofar as the atomic structure of matter can be ignored.

In the atomic picture, thermodynamic quantities are subject to small fluctuations. The second law of thermodynamics is true only on the macroscopic scale, when such fluctuations can be neglected. It is constantly being violated on the atomic scale.

Taking the second law as absolute would lead to the conclusion that the entropy of the universe must forever increase, leading toward an ultimate "heat death." Needless to say, this ceases to be imperative in the atomic picture. We shall discuss the meaning of macroscopic irreversibility in more detail in Section 6.9.

Problems

2.1 An ideal gas undergoes a reversible transformation along the path $P = aV^b$, where a and b are constants, with $a > 0$. Find the heat capacity C along this path.

2.2 The temperature in a lake is 300 K at the surface, and 290 K at the bottom. What is the maximum energy that can be extracted thermally from 1 gr of water by exploiting the temperature difference?

2.3 A nuclear power plant generates 1 MW of power at a reactor temperature of 600°F. It rejects waste heat into a nearby river, with a flow rate of 6000 ft³/s, and an upstream temperature of 70°F. The power plant operates at 60% of maximum possible efficiency. Find the temperature rise of the river.

2.4 A slow-moving stream carrying hot spring water at temperature T_2 joins a sluggish stream of glacial water at temperature T_1. The water downstream has a temperature T between T_2 and T_1, and flows much faster, at a velocity v. The velocities of the input streams can be neglected. Assume that the specific heat of water is a constant, and that the entropy of water has the same temperature dependence as that of an ideal gas.

(a) Find v, neglecting the velocities of the input streams.

Hint: The net change in thermal energy was convert to kinetic energy.

(b) Find the lower bound on T and the upper bound on v imposed by the second law of thermodynamics.

Hint: The total entropy cannot decrease.

2.5 A cylinder of cross-sectional area A is divided into two chambers 1 and 2, by means of a frictionless piston. The piston as well as the walls of the chambers are heat-insulating, and the chambers initially have equal length L. Both chambers are filled with 1 mol of helium gas, with initial pressures $2P_0$, P_0, respectively. The piston is then allowed to slide freely, whereupon the gas in chamber 1 pushes the piston a distance a to equalize the pressures to P.

(a) Find the distance a traveled by the piston.

(b) If W is the work done by gas in chamber 1, what are the final temperatures T_1, T_2 in the two chambers? What is the final pressure P?

(c) Find the work done W.

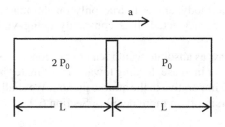

2.6 The equation of state of radiation is $PV = U/3$. Stefan's law gives $U/V = \sigma T^4$, with $\sigma = \pi^2 k^4/(15\hbar^3 c^3)$.

(a) Find the entropy of radiation as a function of V and T.

(b) During the Big Bang, radiation initially confined within a small region expands adiabatically and cools down. Find the relation between the temperature T and the radius of the universe R.

2.7 Put an ideal gas through a Carnot cycle and show that the efficiency is $\eta = 1 - T_1/T_2$, where T_2 and T_1 are the ideal gas temperatures of the heat reservoirs. This shows that the ideal gas temperature coincides with the absolute temperature.

2.8 The Diesel cycle is illustrated in the accompanying diagram. Let $r = V_1/V_2$ (compression ratio), and $r_c = V_3/V_2$ (cutoff ratio). Assuming that the working substance is an ideal gas with $C_P/C_V = \gamma$, find the efficiency of the cycle.

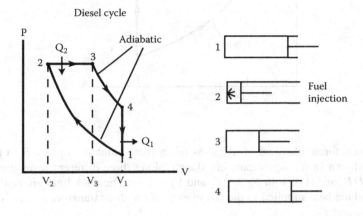

2.9 The Otto cycle, as shown in the accompanying diagram, where $r = V_1/V_2$ is the compression ratio. The working substance is an ideal gas. Find the efficiency.

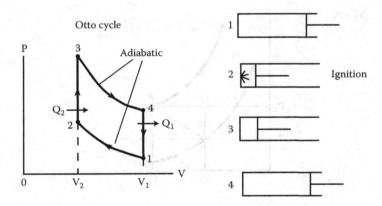

2.10 An ideal gas undergoes a cyclic transformation *abca* (see sketch) such that

> *ab* is at constant pressure, with $V_b = 2V_a$,
> *bc* is at constant volume,
> *ca* is at constant temperature.

Find the efficiency of this cycle, and compare it with that of the Carnot cycle operating between the highest and lowest available temperatures.

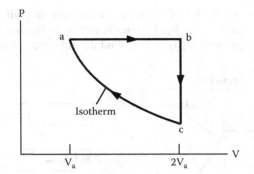

2.11 A monatomic classical ideal gas is taken from point A to point B in the PV diagram shown in the accompanying sketch, along three different reversible paths ACB, ADB, and AB, with $P_2 = 2P_1$ and $V_2 = 2V_1$. The thick lines are isotherms.

(a) Find the heat supplied to the gas in each of the three transformations, in terms of Nk, P_1, and T_1.

(b) What is the heat capacity along the path AB?

(c) Find the efficiency of the heat engine based on the closed cycle $ACBD$.

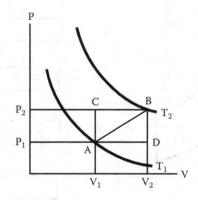

2.12 Here's a device that allegedly violates the second law of thermodynamics. Consider the surface of rotation shown in heavy lines in the accompanying sketch. It is made up of parts of the surfaces of two confocal ellipsoids of revolution, and that of a sphere. The inside surface is a perfect mirror. The foci are labeled A and B.

The argument goes as follows: If two black bodies of equal temperature are initially placed at A and B, then all the radiation from A will reach B, but not vice versa, because radiation from B hitting the spherical surface will be reflected back. Therefore the temperature of B will increase spontaneously, while that of A will decrease spontaneously, and this would violate the second law.

(a) Why would a spontaneous divergence of temperatures violate the second law?
(b) Is the assertion true for physical black bodies?

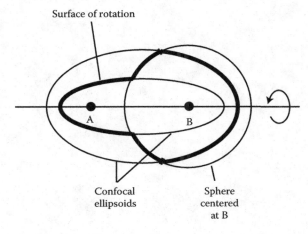

Surface of rotation

A   B

Confocal
ellipsoids

Sphere
centered
at B

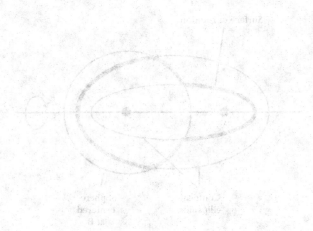

Chapter 3

Using Thermodynamics

3.1   The Energy Equation

In Chapter 2 we obtained the dQ equations [Equation (2.6)], which tell us the amount of heat absorbed by a system when the independent variables change. However, the formulas involve derivatives of the internal energy, which is not directly measurable. We can obtain more practical results by rewriting the equations exploiting the fact that $dS = dQ/T$ is an exact differential. We illustrate the method using T, V as independent variables. From Equation (2.6),

$$dQ = TdS = C_V \, dT + \left[\left(\frac{\partial U}{\partial V} \right)_T + P \right] dV \tag{3.1}$$

Dividing both sides by T, we obtain

$$dS = \frac{C_V}{T} \, dT + \frac{1}{T} \left[\left(\frac{\partial U}{\partial V} \right)_T + P \right] dV \tag{3.2}$$

Being an exact differential, this should be of the form

$$dS = \frac{\partial S}{\partial T} \, dT + \frac{\partial S}{\partial V} dV \tag{3.3}$$

where we suppress the subscripts on partial derivatives for simplicity. Thus, we can identify

$$\frac{\partial S}{\partial T} = \frac{C_V}{T}$$
$$\frac{\partial S}{\partial V} = \frac{1}{T} \left[\frac{\partial U}{\partial V} + P \right] \tag{3.4}$$

Since differentiation is a commutative operation, we have

$$\frac{\partial}{\partial V} \frac{\partial S}{\partial T} = \frac{\partial}{\partial T} \frac{\partial S}{\partial V} \tag{3.5}$$

Hence

$$\frac{\partial}{\partial V} \frac{C_V}{T} = \frac{\partial}{\partial T} \left[\frac{1}{T} \left(\frac{\partial U}{\partial V} + P \right) \right] \tag{3.6}$$

33

The left side can be written as $T^{-1} \partial C_V / \partial V$, since T is kept fixed for the differentiation. Using $C_V = \partial U / \partial T$, we can rewrite the last equation in the form

$$\frac{1}{T} \frac{\partial}{\partial V} \frac{\partial U}{\partial T} = -\frac{1}{T^2} \left[\frac{\partial U}{\partial V} + P \right] + \frac{1}{T} \left[\frac{\partial}{\partial V} \frac{\partial U}{\partial T} + \frac{\partial P}{\partial T} \right] \tag{3.7}$$

After cancelling identical terms on both sides, we obtain the *energy equation*

$$\left(\frac{\partial U}{\partial V} \right)_T = T \left(\frac{\partial P}{\partial T} \right)_V - P \tag{3.8}$$

where we have restored the subscripts. The derivative of the internal energy is now expressed in terms of readily measurable quantities.

In Chapter 2, we deduced from Joule's free expansion experiment that the internal energy for an ideal gas depends only on the temperature, and not the volume, that is, $(\partial U / \partial V)_T = 0$. Now we can show that this is implied by the second law, through the energy equation. For the ideal gas we have

$$\left(\frac{\partial P}{\partial T} \right)_V = \frac{P}{T} \quad \text{(ideal gas)} \tag{3.9}$$

by the equation of state. Therefore, by the energy equation,

$$\left(\frac{\partial U}{\partial V} \right)_T = 0 \quad \text{(ideal gas)} \tag{3.10}$$

3.2   Some Measurable Coefficients

The energy equation relates the experimental inaccessible quantity $(\partial U / \partial V)_T$ to $(\partial P / \partial T)_V$. The latter can be related to other thermodynamic coefficients, using the chain rule for partial derivatives (see Appendix). We can write

$$\left(\frac{\partial P}{\partial T} \right)_V = -\frac{1}{(\partial T / \partial V)_P (\partial V / \partial P)_T} = \frac{(\partial V / \partial T)_P}{-(\partial V / P)_T} = \frac{\alpha}{\kappa_T} \tag{3.11}$$

where α and κ_T are among some directly measurable coefficients:

$$\alpha = \frac{1}{V} \left(\frac{\partial V}{\partial T} \right)_P \quad \text{(coefficient of thermal expansion)}$$

$$\kappa_T = -\frac{1}{V} \left(\frac{\partial V}{\partial P} \right)_T \quad \text{(isothermal compressibility)}$$

$$\kappa_S = -\frac{1}{V} \left(\frac{\partial V}{\partial P} \right)_S \quad \text{(adiabatic compressibility)} \tag{3.12}$$

Substituting the new form of $(\partial P/\partial T)_V$ into $(\partial U/\partial V)_T$ and then the latter into the dQ equation [Equation (3.1)], we obtain

$$TdS = C_V dT + \frac{\alpha T}{\kappa_T} dV \qquad (3.13)$$

This "TdS equation" gives the heat absorbed in terms of directly measurable coefficients. Using T, P as independent variables, we have

$$TdS = C_P dT - \alpha T V dP \qquad (3.14)$$

To use V, P as independent variables, we rewrite dT in terms of dV and dP:

$$dT = \left(\frac{\partial T}{\partial V}\right)_P dV + \left(\frac{\partial T}{\partial P}\right)_V dP = \frac{1}{\alpha V} dV + \frac{\kappa_T}{\alpha} dP \qquad (3.15)$$

In summary the TdS equations are

$$TdS = C_V dT + \frac{\alpha T}{\kappa_T} dV$$

$$TdS = C_P dT - \alpha T V dP$$

$$TdS = \frac{C_P}{\alpha V} dV + \left(\frac{C_P \kappa_T}{\alpha} - \alpha TV\right) dP \qquad (3.16)$$

3.3 Entropy and Loss

In Chapter 2 we showed that the entropy of an isolated system remains constant during a reversible process, but it increases during an irreversible process. Useful energy is lost in the process, and the increase of entropy measures the loss.

To illustrate this point, consider 1 mol of an ideal gas at temperature T, expanding from volume V_1 to V_2. Let us compare the entropy change for a reversible isothermal expansion and an irreversible free expansion. The processes are schematically depicted in Figure 3.1. They have the same initial and final states, as indicated on the PV diagram in Figure 3.2. However, the path for the free expansion cannot be represented on the diagram, because the pressure is not well-defined.

In the reversible isothermal expansion, $\Delta U = 0$ because U depends only on the temperature for an ideal gas. By the first law, we have

$$\Delta Q = \Delta W = \int_{V_1}^{V_2} \frac{RT}{V} dV = RT \ln \frac{V_2}{V_1} \qquad (3.17)$$

The work done ΔW is stored in a spring attached to the moving wall, and it can be used to compress the gas back to the initial state. The entropy change of the gas is

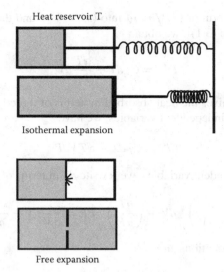

Figure 3.1 Reversible isothermal expansion and irreversible free expansion. In the former case, the temperature is maintained by a heat reservioir, and the work done is stored externally.

given by

$$(\Delta S)_{\text{gas}} = \int \frac{dQ}{T} = \frac{\Delta Q}{T} = R \ln \frac{V_2}{V_1} \qquad (3.18)$$

Since the heat reservoir delivers an amount of heat ΔQ to the gas at temperature T, its entropy change is

$$(\Delta S)_{\text{reservoir}} = -\Delta Q/T = -R \ln \frac{V_2}{V_1} \qquad (3.19)$$

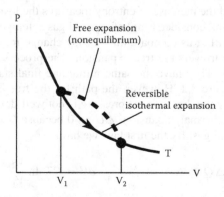

Figure 3.2 Isothermal expansion and free expansion. The latter cannot be represented by a path, because pressure is not well-defined during the process.

Thus, there is no change in the entropy of the "universe" made up of the system and its environment:

$$(\Delta S)_{\text{universe}} \equiv (\Delta S)_{\text{gas}} + (\Delta S)_{\text{reservoir}} = 0 \tag{3.20}$$

In the free expansion, $\Delta W = 0$ because the gas expands into a vacuum. since the temperature of the gas does not change, according to Joule's experiment, there is no heat transfer between the gas and its environment. Therefore

$$(\Delta S)_{\text{reservoir}} = 0 \tag{3.21}$$

The entropy change of the gas is the same as that calculated before, since it only depends on the initial and final states:

$$(\Delta S)_{\text{gas}} = R \ln \frac{V_2}{V_1} \tag{3.22}$$

Therefore

$$(\Delta S)_{\text{universe}} = R \ln \frac{V_2}{V_1} \tag{3.23}$$

Had the transformation proceeded reversibly, the gas could have performed work in the amount

$$\Delta W = RT \ln \frac{V_2}{V_1} = T (\Delta S)_{\text{universe}} \tag{3.24}$$

The increase in total entropy is a reflection of the loss of useful energy.

In heat conduction, an amount of heat ΔQ is directly transferred from a hotter reservoir T_2 to a cooler one T_i, and the entropy change of the universe is

$$(\Delta S)_{\text{universe}} = \frac{\Delta Q}{T_2} - \frac{\Delta Q}{T_1} > 0 \tag{3.25}$$

This shows that heat conduction is always irreversible. The only reversible way to transfer heat from T_2 to T_2 is to connect a Carnot engine between the two reservoirs, so that the work output can be used to reverse the process.

3.4 *TS* Diagram

We can use the entropy S as independent variable and represent a thermodynamic process on a *TS* diagram. In such a representation, adiabatic lines are vertical lines, and the area under a path is the heat absorbed: $\int T dS = \Delta Q$.

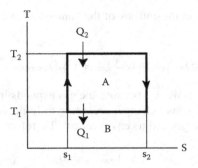

Figure 3.3   Carnot cycle on a *TS* diagram.

A Carnot cycle is a rectangle, as illustrated in Figure 3.3. The heat absorbed in the cycle is equal to the area $A + B$. The heat rejected is the area B, and the total work output is A. The efficiency is therefore

$$\eta = \frac{A}{A + B} \qquad (3.26)$$

The *TS* diagram is helpful in analyzing non-Carnot cycles. As an example, consider the cycle shown in Figure 3.4. The cycle in the left panel is equivalent to a composite cycle made up of the two Carnot cycles 1 and 2 in the right panel. Their efficiencies are given by the corresponding ratios of the areas indicated in Figure 3.4:

$$\eta_1 = \frac{A}{A + B}$$

$$\eta_2 = \frac{C}{C + D} \qquad (3.27)$$

From Figure 3.4 we can see that $\eta_1 > \eta_2$, since cycle 1 has a larger ratio of working temperatures.

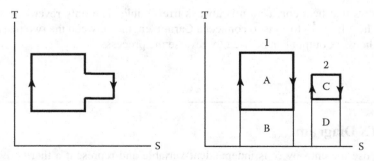

Figure 3.4   The non-Carnot cycle on the left panel is equivalent the two cycles shown on the right panel.

Now the combined cycle has efficiency

$$\eta = \frac{A + C}{A + B + C + D} \tag{3.28}$$

A pure Carnot cycle working between the highest and lowest available temperature will have the efficiency η_1, and we expect that to be more efficient than the composite cycle, that is, $\eta < \eta_1$.

To see this, write

$$\frac{1}{\eta} = \frac{A + B + C + D}{A + C} = \frac{A + B}{A + C} + \frac{C + D}{A + C}$$

$$= \frac{1}{\eta_1} \frac{A}{A + C} + \frac{1}{\eta_2} \frac{C}{A + C} \tag{3.29}$$

Since $\eta_1 > \eta_2$, we can replace η_2 by η_1 in the above, and arrive at the inequality

$$\frac{1}{\eta} > \frac{1}{\eta_1} \frac{A}{A + C} + \frac{1}{\eta_1} \frac{C}{A + C} = \frac{1}{\eta_1} \tag{3.30}$$

Thus

$$\eta < \eta_1 \tag{3.31}$$

3.5 Condition for Equilibrium

The first law states $\Delta U = \Delta Q - P \Delta V$. Using Clausius' theorem $\Delta Q \leq T \Delta S$, we have

$$\Delta U \leq T \Delta S - \Delta W \tag{3.32}$$

Thus, $\Delta U \leq 0$ for a system with $\Delta S = \Delta W = 0$. This means that the internal energy will seek the lowest possible value, when the system is thermally and mechanically isolated. For infinitesimal reversible changes we have

$$dU = T dS - P dV \tag{3.33}$$

Thus, the natural variables for U are S and V. If the function $U(S, V)$ is known, we can obtain all thermodynamic properties through the formulas

$$P = -\left(\frac{\partial U}{\partial V}\right)_S$$

$$T = \left(\frac{\partial U}{\partial S}\right)_V \tag{3.34}$$

These are known as Maxwell relations.

3.6   Helmholtz Free Energy

In the laboratory it is difficult to manipulate S, V, but far easier to change T, V. It is thus natural to ask, "What is the equilibrium condition at constant T, V ?" To answer this question, we go back to the inequality $\Delta U \leq T\Delta S - \Delta W$. If T is kept constant, we can rewrite it in the form

$$\Delta W \leq -\Delta(U - TS) \tag{3.35}$$

If $\Delta W = 0$, then $(U - TS) \leq 0$. This motivates us to define a new thermodynamic function, the *Helmholtz free energy* (or simply *free energy*):

$$A \equiv U - TS \tag{3.36}$$

The earlier inequality now reads

$$\Delta A \leq -\Delta W \tag{3.37}$$

If $\Delta W = 0$, then $\Delta A \leq 0$. The equilibrium condition for a mechanically isolated body at constant temperature is that the free energy be minimum.

For infinitesimal reversible transformations we have $dA = dU - TdS - SdT$. Using the first law, we can reduce this to

$$dA = -PdV - SdT \tag{3.38}$$

If we know the function $A(T, V)$, then all thermodynamic properties can be obtain through the Maxwell relations

$$P = -\left(\frac{\partial A}{\partial V}\right)_T$$

$$S = -\left(\frac{\partial A}{\partial T}\right)_V \tag{3.39}$$

The first of these reduces to the intuitive relation $P = -\partial U/\partial V$ at absolute zero.

To illustrate the minimization of free energy, consider the arrangement shown in Figure 3.5. An ideal gas is contained in a cylinder divided into two compartments of volumes V_1 and V_2, respectively, with a dividing wall that can slide without friction.

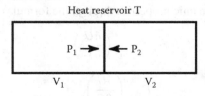

Figure 3.5   The sliding partition will come to rest at such a position as to minimize the free energy. This implies equalization of pressure: $P_1 = P_2$.

The entire system is in thermal contact with a heat reservoir of temperature T. As the partition slides, the total volume $V = V_1 + V_2$ as well as the temperature T remain constant. Intuitively we know that the partition will slide to such a position as to equalize the pressures on both sides. How can we show this purely on thermodynamic grounds? The answer is that we must minimize the free energy. This means that, when equilibrium is established, any small displace of the partition will produce no change in the free energy to first order, that is,

$$\delta A = \frac{\partial A}{\partial V_1} \delta V_1 + \frac{\partial A}{\partial V_2} \delta V_2 = 0 \tag{3.40}$$

with the constraint $\delta V_1 + \delta V_2 = 0$. Thus the condition for equilibrium is

$$\left(\frac{\partial A}{\partial V_1} - \frac{\partial A}{\partial V_2} \right) \delta V_1 = 0 \tag{3.41}$$

Since T is constant, the partial derivatives give the pressures. Hence $P_1 = P_2$.

3.7 Gibbs Potential

We have seen that the thermodynamic properties of a system can be obtained from the function $U(S, V)$, or from $A(V, T)$, depending on the choice of independent variables. The replacement of U by $A = U - TS$ was motivated by the fact that $dU = TdS - PdV$, and we want to replace the term TdS by SdT. This is an example of a *Legendre transformation*.

Let us now consider P, T as independent variables. We introduce the Gibbs potential G, by making a Legendre transformation on A:

$$G \equiv A + PV \tag{3.42}$$

Then, $dG = dA + PdV + VdP = -SdT - PdV + PdV + VdP$, or

$$dG = -SdT + VdP \tag{3.43}$$

The condition for equilibrium at constant T, P is that G be at a minimum. We now have further Maxwell relations

$$V = \left(\frac{\partial G}{\partial P} \right)_T$$

$$S = -\left(\frac{\partial G}{\partial T} \right)_P \tag{3.44}$$

The Gibbs potential is useful in describing chemical processes, which usually take place under constant atmospheric pressure.

Figure 3.6   Mnemonic diagram summarizing the Maxwell relations. Each quantity at the center of a row or column is flanked by its natural variables. The partial derivative with respect to one variable, with the other kept fixed, is arrived at by following the diagonal line originating from that variable. Attach a minus sign if you go against the arrow.

3.8   Maxwell Relations

The following basic functions are related to one other through Legendre transformations:

$$U(S, V): \quad dU = TdS - PdV$$

$$A(V, T): \quad dA = -SdT - PdV$$

$$G(P, T): \quad dG = -SdT + VdP$$

$$H(S, P): \quad dH = TdS - VdP \tag{3.45}$$

Each function is expressed in terms of its natural variables. When these variables are held fixed, the corresponding function is at a minimum in thermal equilibrium. Thermodynamic functions can be obtained through the Maxwell relations summarized in the diagram in Figure 3.6.

3.9   Chemical Potential

So far we have kept the number of particles N constant in thermodynamic transformations. When N does change, the first law is generalized to the form

$$dU = dQ - dW + \mu dN \tag{3.46}$$

where μ is called the *chemical potential,* the energy needed to add one particle to a thermally and mechanically isolated system. For a gas-liquid system we have

$$dU = TdS - PdV + \mu dN \tag{3.47}$$

The change in free energy is given by

$$dA = -SdT - PdV + \mu dN \tag{3.48}$$

which gives the Maxwell relation

$$\mu = \left(\frac{\partial A}{\partial N}\right)_{V,T} \tag{3.49}$$

Similarly, for processes at constant P and T, we consider the change in the Gibbs potential:

$$dG = -SdT - VdP + \mu dN \tag{3.50}$$

and obtain

$$\mu = \left(\frac{\partial G}{\partial N}\right)_{P,T} \tag{3.51}$$

Problems

3.1 We derive some useful thermodynamic relations in this problem.

(a) The TdS equations remain valid when $dS = 0$. Exploiting this fact, express C_V and C_P in terms of adiabatic derivatives, and show

$$\frac{C_P}{C_V} = \frac{\kappa_T}{\kappa_S}$$

(b) Equate the right sides of the first two TdS equations, and then use P, V as independent variables. From this derive the relation

$$C_P - C_V = \frac{\alpha^2 TV}{\kappa_T}$$

(c) Using the Maxwell relations show

$$C_V = -T^2 \left(\frac{\partial^2 A}{\partial T^2}\right)_V$$

$$C_P = -T^2 \left(\frac{\partial^2 G}{\partial T^2}\right)_P$$

3.2 When the number of particles changes in a thermodynamic transformation, it is important to use the correct form of the entropy for an ideal gas, as given by the Sacker–Tetrode equation [Equation (2.50)].

(a) Use the Sacker–Tetrode equation to calculate $A(V, T)$ and $G(P, T)$ for an ideal gas. Show in particular

$$A(V, T) = Nk_B T[\ln(n\lambda^3) - 1]$$

where n is the density, and $\lambda = \sqrt{2\pi\hbar^2/mk_BT}$ is the thermal wavelength.

(b) Obtain the chemical potential for an ideal gas from $(\partial A/\partial N)_{V,T}$ and $(\partial G/\partial N)_{P,T}$. Show that you get the same answer

$$\mu = k_B T \ln(n\lambda^3)$$

3.3 A glass flask with a long narrow neck, of small cross-sectional area A, is filled with 1 mol of a dilute gas with $C_P/C_V = \gamma$, at temperature T. A glass bead of mass m fits snugly into the neck, and can slide along the neck without friction. Find the frequency of small oscillations of the bead about its equilibrium position. This gives a method to measure γ.

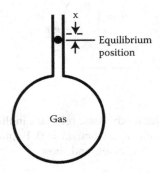

3.4 A cylinder with insulating walls is divided into two equal compartments by means of an insulating piston of mass M, which can slide without friction. The cylinder has cross section A, and the compartments are of length L. Each compartment contains 1 mol of a classical ideal gas with $C_P/C_V = \gamma$, at temperature T_0 (see sketch).

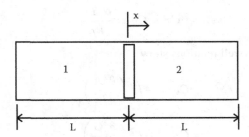

(a) Suppose the piston is adiabatically displaced a small distance $x \ll L$. Calculate, to first order in x, the pressures P_1, P_2 and temperatures T_1, T_2 in the two chambers.

(b) Find the frequency of small adiabatic oscillations of the piston about its equilibrium position.

(c) Now suppose that the piston has a small heat conductivity, so that heat flows from 1 to 2 at the rate $dQ/dt = K\Delta T$, where K is very small, and $\Delta T = T_1 - T_2$. Find the rate of increase of the entropy of the universe.

(d) Entropy generation implies energy dissipation, which damps the oscillation. Calculate the energy dissipated per cycle.

3.5 A liquid has an equilibrium density corresponding to specific volume $v_0(T)$. Its free energy can be represented by

$$A(V, T) = Na_0(T)[v_0(T) - v]^2 - Nf(T)$$

where $v = V/N$.

(a) Find the equation of state $P(v, T)$ of the liquid.

(b) Calculate the isothermal compressibility κ_T and the coefficient of thermal expansion α.

(c) Find the chemical potential.

Note: For $v > v_0$ the pressure becomes negative, and therefore unphysical. See Problem 4.4 for remedy.

3.6 A mixture of two ideal gases undergoes an adiabatic transformation. The gases are labeled 1,2. Their densities and heat capacities are denoted by n_j, C_{Vj}, C_{Pj} $(j = 1, 2)$. Show that the pressure P and volume V of the system obey the constraint $PV^\xi = $ constant, where

$$\xi = \frac{n_1 C_{P1} + n_2 C_{P2}}{n_1 C_{V1} + n_2 C_{V2}}$$

Hint: The entropy of the system does not change, but those of the components do. For the entropy change of an ideal gas, use $\Delta S = Nk \ln[(V_f/V_i)(T_f/T_i)^{3/2}]$, where f and i denote final and initial values.

3.7 Two thin disks of metal were at temperatures T_2 and T_1, respectively, with $T_2 > T_1$. They are brought into thermal contact on their flat surfaces, and came to equilibrium under atmospheric pressure, in thermal isolation.

(a) Find the final temperature.

(b) Find the increase in entropy of the universe.

3.8 The thermodynamic variables for a magnetic system are H, M, T, where H is the magnetic field, M the magnetization, and T the absolute temperature. The magnetic work is $dW = -HdM$, and the first law states $dU = TdS + HdM$. The equation of state is given by Curie's law $M = \kappa H/T$, where $\kappa > 0$ is a constant. This is valid only at small H and high T. The heat capacity at constant H is denoted by C_H, and that at constant M is C_M. Many thermodynamic relations can be obtained from those of a PVT system by the correspondence $H \longleftrightarrow -P, M \longleftrightarrow V$

(a) Show

$$C_M = \left(\frac{\partial U}{\partial T}\right)_M$$

$$C_H = \left(\frac{\partial U}{\partial T}\right)_H - H\left(\frac{\partial M}{\partial T}\right)_H$$

(b) From the analog of the energy equation and Curie's law, show that

$$\left(\frac{\partial U}{\partial M}\right)_T = 0$$

This is the analog of the statement that the internal energy of the ideal gas is independent of the volume.

(c) Show

$$C_H - C_M = \frac{M^2}{\kappa}$$

3.9 Define the free energy $A(M, T)$ and Gibbs potential $G(H, T)$ of a magnetic system by analogy with the PVT system.

(a) Show

$$dA = -SdT + HdM$$

$$dG = -SdT - MdH$$

(b) Show $(\partial S/\partial H)_T = (\partial M/\partial T)_H$, hence

$$\left(\frac{\partial S}{\partial H} \right)_T = -\frac{\kappa H}{T^2}$$

(b) With the help of the chain rule, show

$$\left(\frac{\partial T}{\partial H} \right)_S = \frac{\kappa H}{C_H T}$$

3.10 Adiabatic demagnetization A paramagnet cools when the magnetic field is decreased adiabatically. The path shown in the accompanying diagram can be used to cool the system to very low temperatures.

(a) Using the properties derived in the last problem, verify the qualitative behavior of S as a function of T and H.

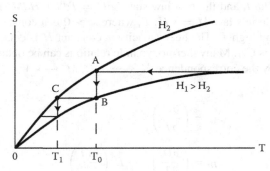

(b) Put $H_2 = 0$, $H_1 = H$. Assume $C_M = aT^3$, where a is a constant. Find the heat absorbed by the system along the path of isothermal magnetization $A \rightarrow B$.

(c) In the adiabatic demagnetization $B \rightarrow C$, calculate T_1 as a function of T_0 and H.

Chapter 4

Phase Transitions

4.1 First-Order Phase Transition

When water boils, it undergoes a phase transition from a liquid to a gas phase. The equation of state in each of these phases is a regular function, continuous with continuous derivatives; but in going from one phase to another it abruptly changes to a different regular function. The two phases can coexist in equilibrium at given pressure and temperature. This means that the Gibbs potential should be continuous as we go from one phase to the others. However, since the two phases differ in density and specific entropy, the first derivatives of the Gibbs potential are discontinuous (see Equation [4.6]). For this reason, the gas-liquid transition is called a "first-order phase transition."

There are second-order phase transitions, in which the first derivatives are continuous, but the second derivatives are discontinuous. Examples include the gas-liquid transition at the critical point, the ferromagnetic transition, and the superconducting transition. In this chapter, we shall discuss the first-order transition in some detail. The second-order phase transition will be discussed in Chapter 19.

An isotherm exhibiting a first-order gas-liquid transition is shown in the PV diagram of Figure 4.1. At point 1 the system is all liquid, at point 2 it is all gas, and in between the system is a mixture of liquid and gas in the states 1 and 2, respectively. They coexist in thermal equilibrium, at a pressure called the *vapor pressure*. Since the two phases have different densities, the total volume changes at constant pressure as the relative proportion is change, generating the horizontal portion of the isotherm.

The PV diagram is a projection of the equation-of-state surface shown in Figure 4.2, where we have included a first-order liquid-solid phase transition as well. The gas-liquid transition region is topped by a critical point, which lies on the critical isotherm. The transition regions are ruled surfaces perpendicular to the pressure axis. When projected onto the PT plane, they appear as lines representing phase boundaries. This is depicted in Figure 4.3.

Following are some important properties:

- **Latent heat**: Since the coexisting phases have different entropies, the system must absorb or release heat during a phase transition. When a unit amount of phase 1 is converted into phase 2, an amount of *latent heat* is liberated:

$$l = T_0(s_2 - s_1) \qquad (4.1)$$

47

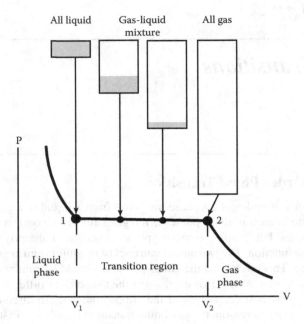

Figure 4.1   An isotherm exhibiting a first-order phase transition.

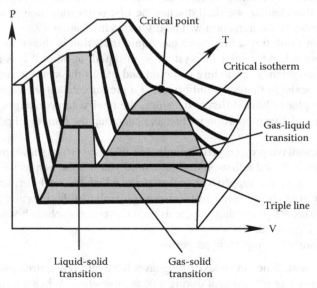

Figure 4.2   Equation of state surface, showing gas-liquid and liquid-solid phase transitions, both of first order. (Not to scale.) Isotherms are shown as heavy lines and the transitions regions are shaded gray.

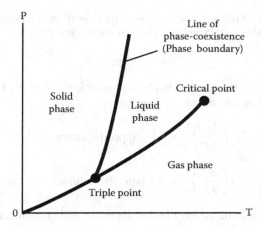

Figure 4.3 The *PT* diagram shows phase boundaries or lines of phase coexistence.

where T_0 is the absolute temperature at which the two phases can coexist, and s_i is the specific entropy of the ith phase. The specific entropy may mean entropy per particle, per mole, per unit mass, or per unit volume.

• **Critical point**: At the *critical point* the gas and the liquid have equal density and specific entropy. The phase transition becomes second-order at this point. There is no critical point for the liquid-solid transition.

• **Triple point**: The line along which gas, liquid, and solid can all coexist projects onto the *triple point* in the *PT* diagram.

4.2 Condition for Phase Coexistence

In a first-order transition, the coexisting phases have the same P, T. The condition for equilibrium, therefore, is that the total Gibbs potential be minimum. The relative proportion of the two phases present may vary, but each phase is characterized by its chemical potential, the Gibbs potential per particle. We use the equivalent quantity $g_i(P, T)$, the Gibbs potential per unit mass, and represent the total Gibbs potential in the form

$$G = m_1 g_1 + m_2 g_2 \qquad (4.2)$$

where m_i is the mass of phase i, with a fixed total mass $m = m_1 + m_2$. The contributions from the two phases are additive because we have neglected surface effects. In equilibrium, G is at a minimum with respect to mass transfer from one phase to the other. If we transfer a small amount of mass δm from one phase to the other at constant P, T, the change δG should vanish to first order:

$$\delta G = g_1 \delta m_1 + g_2 \delta m_2 = 0 \qquad (4.3)$$

with the constraint $\delta m_1 = -\delta m_2 = \delta m$. The condition for phase coexistence is therefore $(g_1 - g_2)\delta m = 0$. Since δm is arbitrary, we must have

$$\Delta g \equiv g_2 - g_1 = 0 \tag{4.4}$$

That is, the coexisting phases must have equal chemical potential. The derivatives of g are discontinuous across the phase boundary:

$$\left(\frac{\partial g}{\partial T}\right)_P = -s \quad \text{(specific entropy)}$$

$$\left(\frac{\partial g}{\partial P}\right)_T = v \quad \text{(specific volume)} \tag{4.5}$$

As the name first-order transition implies, there are finite differences in the specific entropy and specific volume:

$$\Delta s \equiv s_2 - s_1 = -\frac{\partial(\Delta g)}{\partial T} > 0$$

$$\Delta v \equiv v_2 - v_1 = \frac{\partial(\Delta g)}{\partial P} > 0 \tag{4.6}$$

4.3 Clapeyron Equation

Since the three variables $\Delta g, \Delta s, \Delta v$ are functions of P, T, there must exist a relation among them $f(\Delta g, \Delta s, \Delta v) = 0$, and we can apply the chain rule in the form

$$\left(\frac{\partial(\Delta g)}{\partial T}\right)_P \left(\frac{\partial T}{\partial P}\right)_{\Delta g} \left(\frac{\partial P}{\partial(\Delta g)}\right)_T = -1 \tag{4.7}$$

or

$$\frac{(\partial(\Delta g)/\partial T)_P}{(\partial(\Delta g)/\partial P)_T} = -\left(\frac{\partial P}{\partial T}\right)_{\Delta g} \tag{4.8}$$

In equilibrium, $\Delta g = 0$, and $(\partial P/\partial T)_{\Delta g=0}$ gives the slope of the transition line on the PT diagram:

$$\frac{dP}{dT} \equiv \left(\frac{\partial P}{\partial T}\right)_{\Delta g=0} \tag{4.9}$$

For the gas-liquid transition, P is the *vapor pressure*. Substituting this definition into Equation (4.7), and using Equation (4.6), we obtain

$$\frac{dP}{dT} = \frac{\Delta s}{\Delta v} \tag{4.10}$$

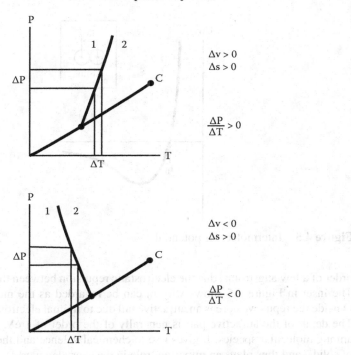

Figure 4.4   The slope of the liquid-solid boundary can have either sign, depending on whether the liquid contracts or expands upon freezing.

or

$$\frac{dP}{dT} = \frac{l}{T\,\Delta v} \qquad (4.11)$$

where $l = T\,\Delta s$ is the latent heat. This is called the *Clapeyron equation*. Depending on the sign of Δv, the slope dP/dT may be positive or negative, as illustrated in Figure 4.4. The upper panel shows the PT diagram for a substance like CO_2, which contracts upon freezing. The lower panel shows that for H_2O, which expands upon freezing.

In a second-order phase transition the first derivatives of g vanish and the Clapeyron equation is replaced by a condition involving second derivatives. (See Problem 4.10.)

4.4   Van der Waals Equation of State

The gas-liquid phase transition owes its existence to intermolecular interactions. The potential energy $U(r)$ between two molecules as a function of their separation r has the qualitative form shown in Figure 4.5. It has a repulsive core whose radius r_0 is of the

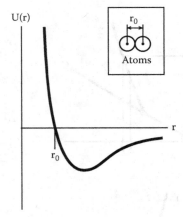

Figure 4.5 Intermolecular potential.

order of a few angstroms, due the electrostatic repulsion between the electron clouds. The inset in Figure 4.5 shows why r_0 can be regarded as the molecular diameter. Outside the repulsive core is an attractive tail due to mutual electrostatic polarization. The depth of the attractive part is generally of the order of 1 eV, but varies widely among molecular species. It gives rise to chemical valence and the crystal structure of solids, and thus plays an important role in the everyday world.

To take the intermolecular interaction into account in a qualitative fashion, we separate the effects of the repulsive core and the attractive tail. The hard core excludes a certain volume around a molecule, so other molecules have less room to move in. The effective volume is therefore smaller than the actual volume:

$$V_{\text{eff}} = V - b \tag{4.12}$$

where V is the total volume of the system, and b is the total excluded volume, of the order of $b \approx N\pi r_0^3/6$. For a fixed number of atoms, it is a constant parameter.

The pressure of gas arises from molecules striking the wall of the container. Compared with the case of the ideal gas, a molecule in a real gas hits the wall with less kinetic energy, because it is being held back by the attraction of neighboring molecules. The reduction in the pressure is proportional to the number of pairs of interacting molecules near the wall, and thus to the density. Accordingly we put

$$P = P_{\text{kinetic}} - \frac{a}{V^2} \tag{4.13}$$

where P_{kinetic} is the would-be pressure in the absence of attraction, and a is a constant proportional to N^2. Van der Waals makes the assumption that, for 1 mole of gas,

$$P_{\text{kinetic}} V_{\text{eff}} = RT \tag{4.14}$$

This leads to

$$(V - b)\left(P + \frac{a}{V^2}\right) = RT \tag{4.15}$$

the *van der Waals equation of state*. In this simple model, the substance is characterized by only two parameters a and b.

4.5   Virial Expansion

The van der Waals equation of state approaches that of the ideal gas in the low density limit: $V \to \infty$ at fixed N. The successive corrections to the ideal gas equation can be obtained by expanding in powers of V^{-1}. We first solve for the pressure:

$$\left(P + \frac{a}{V^2} \right) = \frac{RT}{(V - b)}$$

$$P = \frac{RT}{V} \left(1 - \frac{b}{V} \right)^{-1} + \frac{a}{V^2}$$

$$\frac{PV}{RT} = \left(1 - \frac{b}{V} \right)^{-1} - \frac{a}{RTV} \tag{4.16}$$

Expanding the right side, we obtain

$$\frac{PV}{RT} = 1 + \frac{1}{V} \left(b - \frac{a}{RT} \right) + \left(\frac{b}{V} \right)^2 + \left(\frac{b}{V} \right)^3 + \cdots \tag{4.17}$$

This is of the form of a *virial expansion*:

$$\frac{PV}{RT} = 1 + \frac{c_2}{V} + \frac{c_3}{V^2} + \cdots \tag{4.18}$$

where c_n is called the nth *virial coefficient*. The *second virial coefficient*

$$c_2 = b - \frac{a}{RT} \tag{4.19}$$

can be obtained experimentally by observing deviations from the ideal gas law. By measuring this as a function of T, we can extract the molecular parameters a and b.

4.6   Critical Point

The van der Waals isotherms are sketched in Figure 4.6. The pressure is a cubic polynomial in V:

$$(V - b)(PV^2 + a) = RTV^2$$

$$PV^3 - (bP + RT)V^2 + aV - ba = 0 \tag{4.20}$$

Phase Transitions

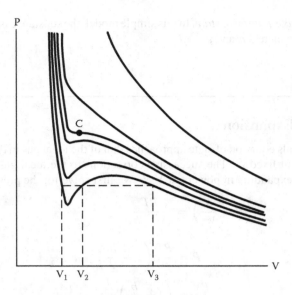

Figure 4.6   Isotherms of the van der Waals gas. The critical point is at C.

There is a region in which the polynomial has three real roots. As we increase T these roots move closer together, and merge at $T = T_c$, the *critical point*. For $T > T_c$, one real root remains, while the other two become a complex-conjugate pair. We can find the critical parameters P_c, V_c, T_c, as follows. At the critical point the equation of state must be of the form

$$(V - V_c)^3 = 0$$

$$V^3 - 3V_c V^2 + 3V_c^2 V - V_c^3 = 0 \qquad (4.21)$$

Comparison with Equation (4.20) yields

$$3V_c = b + \frac{RT_c}{P_c}$$

$$3V_c^2 = \frac{a}{P_c}$$

$$V_c^3 = \frac{ba}{P_c} \qquad (4.22)$$

They can be solved to give

$$RT_c = \frac{8a}{27b}$$

$$P_c = \frac{a}{27b^2}$$

$$V_c = 3b \qquad (4.23)$$

4.7 Maxwell Construction

The van der Waals isotherm is a monotonic function of V for $T > T_c$. Below T_c, however, there is a "kink" exhibiting negative compressibility. This is unphysical, and its origin can be traced to the implicit assumption that the density is always uniform. Actually, as we shall see, the system prefers to undergo a first-order phase transition, by breaking up into a mixture of phases of different densities.

According to the Maxwell relation $P = -(\partial A/\partial V)_T$, the free energy can be obtained as the area under the isotherm:

$$A(V, T) = - \int_{\text{isotherm}} PdV \qquad (4.24)$$

Let us carry out the integration graphically, as indicated in Figure 4.7. The volumes V_1, V_2 are defined by the double-tangent construction. At any point along the tangent, such as X, the free energy is a linear combination of those at 1 and 2, and thus represents a mixture of two phases. This nonuniform state has the same P and T as the uniform state 3, but it has a lower free energy, as is obvious from the graphical construction. Therefore the phase-separated state is the equilibrium state.

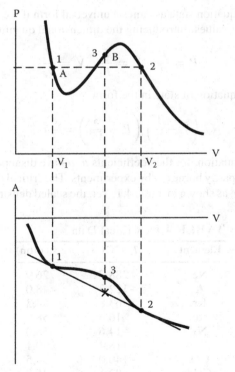

Figure 4.7 The Maxwell construction.

The states 1 and 2 are defined by the conditions

$$\frac{\partial A}{\partial V_1} = \frac{\partial A}{\partial V_2} \quad \text{(equal pressure)}$$

$$\frac{A_2 - A_1}{V_2 - V_1} = \frac{\partial A}{\partial V_1} \quad \text{(common tangent)} \quad (4.25)$$

Thus,

$$-(A_2 - A_1) = -\frac{\partial A}{\partial V_1}(V_2 - V_1)$$

$$\int_{V_1}^{V_2} P dV = P_1(V_2 - V_1) \quad (4.26)$$

This means the areas A and B in Figure 4.7 are equal to each other. The horizontal line in Figure 4.7 is known as the *Maxwell construction*.

4.8 Scaling

The van der Waals equation state assumes a universal form if we measure P, V, T in terms of their critical values. Introducing the dimensional quantities

$$\bar{P} = \frac{P}{P_c} \quad \bar{T} = \frac{T}{T_c} \quad \bar{V} = \frac{V}{V_c} \quad (4.27)$$

we can rewrite the equation of state in the form

$$\left(\bar{V} - \frac{1}{3}\right)\left(\bar{P} + \frac{3}{\bar{V}^2}\right) = \frac{8}{3}\bar{T} \quad (4.28)$$

This is a universal equation, for the coefficients a, b have disappeared.

The scaling law is partly borne out by experiments. The critical data for the elements involved vary widely, as shown in Table 4.1. Yet, the scaled densities of the liquid and

TABLE 4.1 Critical Data

Element	T_c (°C)	P_c (atm)
Ne	−288.7	26.9
A	−122.3	48.0
Kr	−63.8	54.3
Xe	16.6	58.0
N_2	−14.6	58.0
O_2	−118.4	50.1
CO	−140.0	34.5
CH_4	−82.1	45.8

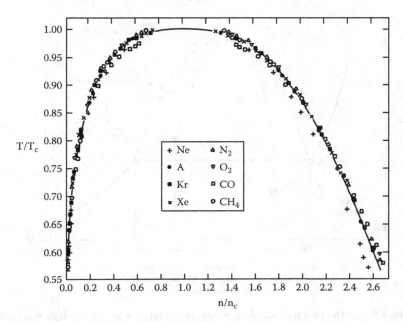

Figure 4.8   The boundary curve of the gas-liquid transition region becomes universal when the temperature T and the density n are measured in term of their values at the critical point. (After Guggenheim 1945.)

gas at the phase boundary depend on the scaled temperature in a universal manner, as shown in Figure 4.8. The universal behavior can be fit by

$$\frac{n_L - n_G}{n_c} = \frac{7}{2}\left(1 - \frac{T}{T_c}\right)^{1/3} \tag{4.29}$$

where n_L is the density of the liquid phase, and n_G is that of the gas phase. The exponent 1/3 is a "critical exponent" conventionally denoted by the symbol β. Although the van der Waals equation of state predicts the scaling behavior, it gives $\beta = 1/2$ instead of the experimental value (Huang 1987). This indicates that the model is not adequate for quantitative purposes.

4.9   Nucleation and Spinodal Decomposition

The van der Waals equation of state is derived under the implicit assumption that the entire system has uniform density. The Maxwell construction in Figure 4.7 shows that, in the transition region, the system can lower its free energy by breaking up into two coexisting phases with different densities. Does the original van der Waals isotherm still have meaning in this region?

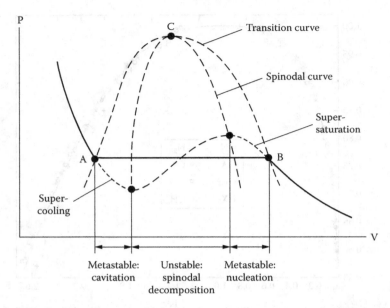

Figure 4.9   In the two metastable regions, new phases emerge through nucleation. In the unstable region, two-phase separation occurs locally and spontaneously, in a process called spinodal decomposition.

It does.

It corresponds to an initial situation that is not in equilibrium. The system was prepared in a state of uniform density, for example through rapid compression. The system needs time to break up into the equilibrium mixture, and the kinetics of this process depends on the compressibility, or curvature of the free energy.

The region in question is shown in the PV diagram of Figure 4.9 as the isotherm between points A and B. Where the compressibility is positive ($\partial P/\partial V < 0$), the system is metastable, and the emergence of a new phase is triggered by nucleation. Where the compressibility is negative ($\partial P/\partial V > 0$), on the other hand, the system is unstable; it breaks up locally and spontaneously, in a process called "spinodal decomposition."

For a closer look we refer to Figure 4.10, and note the following:

- In Figure 4.10, point A lies in a region where the free energy has a positive curvature. The inset shows the local free energy. The local Maxwell construction shows that the uniform state is stable. However, since a global phase separation does lower the free energy, the situation is only metastable. Nucleation triggers phase separation, as illustrated in Figure 4.11. Thermal fluctuations create a microscopic specks of the other phase, of sufficiently large size that it lowers the local free energy. This is the nucleus that eventually grows to macroscopic size.

- Point B in Figure 4.10 lies in a region where the free energy has a negative curvature. The separated state now has lower free energy than the uniform state, and

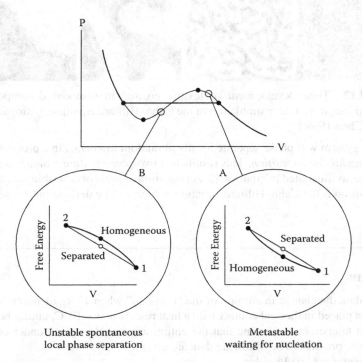

Figure 4.10 In the region A, the free energy has a positive curvature, as shown in the inset; the system is metastable and waits for nucleation of a new phase. In region B, the free energy has a negative curvature; the system is unstable and spontaneously phase separates locally.

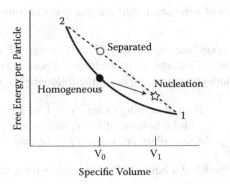

Figure 4.11 When the local free energy has positive curvature, nucleation of a new phase lowers the free energy. The nucleus then grows to macroscopic size.

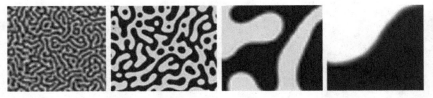

Figure 4.12 Time development of texture created in spinodal decomposition, obtained through computer simulation of the Cahn–Hilliard equation. (Adopted from Zhu and Chen 1999.)

the system will phase separate locally at random moments, in a process called *spinodal decomposition*. This results in a two-phase texture which coarsens in time, as illustrated in Figure 4.12. An equation capable of describing the decomposition is the Cahn–Hilliard equation, which will be derived in Chapter 19.

Problems

4.1 Calculate the change in entropy of the "universe" when 10 kg of water, initially at 20°C, is placed in thermal contact with a heat reservoir at −10°C, until it becomes ice at that temperature. Assume that the entire process takes place under constant atmospheric pressure. The following data are given:

C_P of water = 4180 J/kg-deg

C_P of ice = 2090 J/kg-deg

Heat of fusion for ice = 3.34×10^5 J/kg

4.2 Integrate the Clapeyron equations near a triple point to obtain the equations for the three transition lines meeting at that point. Make the following assumptions:

The gas can be treated as ideal.

The latent heats are constants.

The molar volumes of solid and liquid are nearly equal constants, negligible compared to that of gas.

4.3 Water expands upon freezing, so $dP/dT < 0$ on the PT diagram. (See Figure 4.4.) Calculate the rate of change of the melting temperature of ice with respect to pressure from the Clapeyron equation, using the following data:

$$\text{Heat of melting of ice} \quad \ell = 1.44 \text{ J}$$
$$\text{Molar volume of ice} \quad v_1 = 20 \text{ cm}^3$$
$$\text{Molar volume of water} \quad v_2 = 18 \text{ cm}^3$$

4.4 Reconsider the model of a liquid in Problem 3.5, with free energy

$$A(V, T) = Na_0(T)[v_0(T) - v]^2 - Nf(T)$$

where $v = V/N$, and $v_0(T)$, $a_0(T)$, $f(T)$ are functions of the temperature only.

(a) If $v > v_0(T)$, then the pressure would be negative, assuming that the liquid is being stretched uniformly. Actually, the liquid would rather not fill the entire volume available, but remain at specific volume v_0. Thus, the pressure would remain zero for $v > v_0$. Show this using an argument similar to that for the Maxwell construction. (See accompanying sketch.)

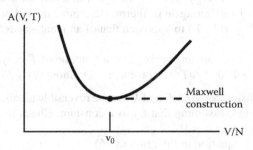

(b) Suppose the liquid is in thermal equilibrium with its vapor, which can be treated as an ideal gas of density n. There are two conditions to be fulfilled, the pressures and chemical potentials must equalize. Write down these conditions. Show $v < v_0$, and $v \to v_0$ only at $T = 0$.

(c) For $nk_BT/a_0 \ll 1$, show $n\lambda^3 = e^{-f/k_BT}$, where $\lambda = \sqrt{2\pi\hbar^2/mk}$.

4.5 (a) In a liquid-gas transition, the specific volume of the liquid (phase 1) is usually negligible compared with that of the gas (phase 2), which usually can be treated as an ideal gas. Let $\ell = T(s_2 - s_1)$ be the latent heat of evaporation per particle. Under the approximations mentioned, show

$$\frac{T}{P}\frac{dP}{dT} = \frac{\ell}{k_BT}$$

(b) Use this formula to obtain the latent heat per unit mass for liquid ^{3}He, in 0.2 K increments of T, from the following table of vapor pressures of ^{3}He. The mass of a ^{3}He atom is $m = 5.007 \times 10^{24}$ g.

T (K)	P (microns of Hg)
0.200	0.0121
0.201	0.0130
0.400	28.12
0.401	28.71
0.600	544.5
0.601	550.3
0.800	2892
0.801	2912
1.000	8842
1.005	9053
1.200	20163
1.205	20529

4.6 Sketch the Gibbs potential $G(P, T)$ of the van der Waals gas as a function of P at constant T. In particular, show the behavior in the transition region. Derive the Maxwell construction using the principle of minimization of G

4.7
  (a) Calculate the free energy $A(V, T)$ for 1 mol of a van der Waals gas.
  Hint: Integrate $-\int P dV$ along an isotherm. Determine the unknown additive function of T by requiring $A(V, T)$ to approach that of an ideal gas as $V \to \infty$, given in Problem 3.2.
  (b) Show that C_V of a van der Waals gas is a function of T only.
  Hint: Use $C_V = -T(\partial^2 A/\partial T^2)_V$ (Problem 3.1). Show $(\partial C_V/\partial V)_T = 0$.

4.8 Find the relationship between T and V for the reversible adiabatic transformation of a van der Waals gas, assuming that C_V is a constant. Check that it reduces to the ideal gas result when $a = b = 0$.
  Hint: Use the TdS equation in Equation (3.16).

4.9 Form the attached table of the second virial coefficient for 1 mol of Ne, find the best fits for the coefficients a and b in the van der Waals equation of state. (Data From Holborn and Otto 1925.)

T (K)	c_2 (cm^3/mol)
60	−20
90	−8
125	0
175	7
225	9
275	11
375	12
475	13
575	14
675	14

4.10 The transition line of a second-order phase transition is shown in the accompanying PT diagram. The first derivatives of the Gibbs potential, giving specific density and entropy, are continuous across this line. The second derivatives, such as compressibilities, are discontinuous. Show that the slope of the transition line is given by

$$\frac{dP}{dT} = \frac{\alpha_1 - \alpha_2}{\kappa_{T1} - \kappa_{T2}}$$

where α denotes the coefficient of thermal expansion, and κ_T the isothermal compressibility.

Hint: Calculate the change of volume ΔV along the transition line. The results for phase 1 and phase 2 must agree.

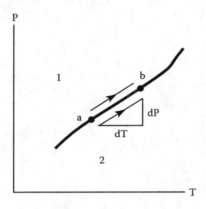

References

Guggenheim, E.A., *J. Chem. Phys.*, **13**:253 (1945).

Holborn and Otto, *Z. Physik*, **33**:1 (1925).

Huang, K., *Statistical Mechanics*, 2nd ed., Wiley, New York, 1987, Section 17.5.

Zhu, J. and L.-Q. *Chen, Phys. Rev.,* **E69**:3564 (1999).

Chapter 5

The Statistical Approach

5.1 The Atomic View

Experiments show that 1 g mol of any dilute gas occupies the same *molar volume*

$$V_0 = 2.24 \times 10^4 \text{ cm}^3 \qquad (5.1)$$

at STP (standard temperature and pressure):

$$T_0 = 273.15 \text{ K}$$

$$P_0 = 1 \text{ atm} \qquad (5.2)$$

From this we can obtain the gas constant $R = P_0 V_0 / T_0$. The ratio of the molar volume to Avogadro's number gives the density of any gas at STP:

$$\text{Density} = 2.70 \times 10^{19} \text{ atoms /cm}^3 \qquad (5.3)$$

This indicates the large number of atoms present in a macroscopic volume. We use the term "atom" here in a generic sense, to denote the smallest unit in the gas, which may in fact be a diatomic molecule such as H_2.

A gas can approach thermal equilibrium because of atomic collisions. The scattering cross section between atoms is of the order of

$$\sigma = \pi r_0^2$$

$$r_0 \approx 10^{-8} \text{cm} \qquad (5.4)$$

where r_0 is the effective atomic diameter. The average distance traveled by an atom between two successive collisions is called the *mean free path*, whose order of magnitude is given by

$$\lambda \approx \frac{1}{n\sigma} \qquad (5.5)$$

where n is the particle density. This can be seen as follows. Consider a cylindrical volume inside the gas, of cross sectional area A, as depicted in Figure 5.1. If the length of the cylinder is small, it contains few atoms, and in the projected end view the images of the atoms rarely overlap. As the length of the cylinder increases, more

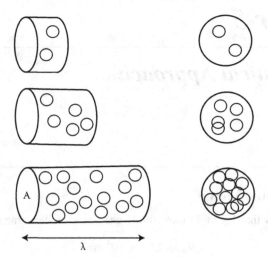

Figure 5.1   The atoms in the project view begin to overlap when the length of the cylinder is of the order of the mean free path λ.

and more atoms are included, and the images begin to overlap when the length of the cylinder is of the order of the mean free path. At that point the area A is roughly equal to the cross-sectional area of all the atoms: $A = (A\lambda)(n\sigma)$, which gives Equation (5.5). At STP we have

$$\lambda \approx 10^{-5} \, \text{cm} \tag{5.6}$$

Thus, on a macroscopic scale, an atom can hardly move without colliding with another atom.

We have learned from thermodynamics that the internal energy per particle in an ideal gas is $\frac{3}{2}k_B T$. Equating this to the average kinetic energy of an atom, we obtain

$$\frac{1}{2}mv^2 = \frac{3}{2}k_B T \tag{5.7}$$

Thus the average velocity is

$$v = \sqrt{\frac{3k_B T}{m}} \tag{5.8}$$

For numerical calculations, it is useful to remember that

$$k_B T \approx \frac{1}{40} \, \text{eV} \quad \text{at } T = 300 \, \text{K}$$

$$m_e c^2 \approx 0.5 \, \text{MeV} \tag{5.9}$$

where m_e is the mass of the electron, and c is the velocity of light. Thus, for H_2 gas at 300 K, we have

$$\frac{v^2}{c^2} = \frac{3k_B T}{mc^2} \approx 3\left(\frac{1}{40}\text{ eV}\right)\frac{1}{0.5 \times 10^6 \times 2 \times 2000 \text{ eV}} \approx 4 \times 10^{-11} \qquad (5.10)$$

Using, $c = 3 \times 10^{10}$ cm/s, we get

$$v \approx 10^5 \text{ cm/s} \qquad (5.11)$$

This is the speed with which a gas expands into a vacuum in a free expansion. The average time between successive collisions is called the *collision time*:

$$\tau = \frac{\lambda}{v} \approx 10^{-10} \text{ s} \qquad (5.12)$$

which is of the order of the relaxation time toward local thermal equilibrium.

5.2 Random Walk

An atom in a gas at STP undergoes 10^{10} collisions per second, on average. We can imagine how complicated its trajectory must be. It would be so tortuous and irregular as to appear random, that is, unpredictable. But thermodynamic order emerges precisely from randomness. To see how disorder can generate order, let us look at a simple model, the random walk.

Consider a particle moving in one dimension, say along the x axis. At regular time intervals it makes steps of equal size, choosing at random whether to go forward or backward. In a walk consisting of n steps, the particle will end up at different positions in different trials, and after a large number of runs there will result a distribution of end positions. Our aim is to calculate that.

Let $W(k, n)$ be the probability that, after taking n random steps, the particle is k steps from the starting point, where $-n \leq k \leq n$. Suppose in one run the particle makes F steps forward and B steps backward. We must have

$$F + B = n$$

$$F - B = k \qquad (5.13)$$

Thus

$$F = (n + k)/2$$

$$B = (n - k)/2 \qquad (5.14)$$

Since each step involves a two-valued decision forward or backward, the total number of possible trajectories is 2^n. Among these, those having F forward steps is just the number of ways to choose the F forward step from n, which is

$$\binom{n}{F} = \frac{n!}{F!(n-F)!} \tag{5.15}$$

Thus

$$W(k, n) = 2^{-n}\binom{n}{F} = \frac{2^{-n}n!}{[(n+k)/2]![(n-k)/2]!} \tag{5.16}$$

For $n \gg 1$, and $n \gg |k|$, we can use the Stirling approximation

$$\ln n! \approx n \ln n - n + \ln \sqrt{2\pi n} \tag{5.17}$$

to obtain[1]

$$W(k, n) \approx \frac{1}{\sqrt{2\pi n}} \exp\left(-\frac{k^2}{2n}\right) \tag{5.18}$$

The probability of returning to the origin after a large number of steps n is obtained by setting $k = 0$:

$$W(0, n) \approx (2\pi n)^{-1/2} \tag{5.19}$$

The model does not contain intrinsic scales for distance and time. To measure in our units, for example centimeter and second, we have to introduce then into the model.

Suppose in our units x_0 is the size of a step, and t_0 is the duration of a step. Making n random steps then corresponds to traveling a total distance x in time t, with

$$x = nx_0$$
$$t = nt_0 \tag{5.20}$$

The probability that the particle ends up between x and $x + dx$ at time t is denoted by $W(x, t)dx$, with

$$W(x, t) = \frac{1}{\sqrt{4\pi Dt}} \exp\left(-\frac{x^2}{4Dt}\right)$$
$$D = \frac{x_0^2}{2t_0} \tag{5.21}$$

The normalization condition is

$$\int_{-\infty}^{\infty} dx W(x, t) = 1 \tag{5.22}$$

---

[1]Calculational note: To get the factor in front, it is necessary to keep the last term in the Sterling approximation $\ln \sqrt{2\pi n}$, and this makes the calculation laborious. However, one can make a short cut by dropping this term and obtain the constant by noting that; since W is a probability, it should be so normalized that $\int_0^\infty dk W = 1$.

The mean-square distance traveled after time t is

$$\langle x^2 \rangle = \int_{-\infty}^{\infty} dx \, x^2 W(x, t) = 4Dt$$

This describes *diffusion*, in which the average distance travel over a time interval t is proportional to $\sqrt{t}$. The constant D is called the *diffusion constant*.

If the random walk takes place in three dimensions, with steps taken independently along the x, y, or z axis, the probability distribution is the product of the three one-dimensional distributions. Because of the homogeneity of space, it depends only on $r = \sqrt{x^2 + y^2 + z^3}$:

$$W(r, t) = \frac{1}{(4\pi D t)^{3/2}} \exp\left(-\frac{r^2}{4Dt}\right) \tag{5.23}$$

The normalization condition is

$$4\pi \int_0^{\infty} dr \, r^2 W(r, t) = 1 \tag{5.24}$$

5.3 Phase Space

To formulate a statistical approach to the classical gas, let us first review how we describe it in classical mechanics. The state of an atom at any instant of time is specified by its position $\mathbf{r}$ and momentum $\mathbf{p}$. The six components of these vector quantities span the phase space of one atom. For N atoms, the total number of degrees of freedom is $6N$, and the total phase space is a $6N$-dimensional space. The motions of the particles are governed by the Hamiltonian

$$H(p, r) = \sum_{i=1}^{N} \frac{\mathbf{p}_i^2}{2m} + \frac{1}{2} \sum_{i \neq j} U(\mathbf{r}_i - \mathbf{r}_j) \tag{5.25}$$

where $U(\mathbf{r})$ is the interatomic potential. The Hamiltonian equations of motions are

$$\dot{\mathbf{p}}_i = \frac{\partial H}{\partial \mathbf{r}_i}$$

$$\dot{\mathbf{r}}_i = -\frac{\partial H}{\partial \mathbf{p}_i} \tag{5.26}$$

In the absence of external time-dependent forces, H has no explicit dependence on the time. The value of the Hamiltonian is the energy, a constant of the motion.

We shall use the shorthand (p, r) to denote all the momenta and coordinates. The $6N$-dimensional space spanned by (p, r) is called the Γ-space. A point in this space, called a *representative point*, corresponds to a state of the N-body system at a particular time. As time evolves, the representative point traces out a trajectory. It never intersects itself, because the solution to the equations of motion is unique,

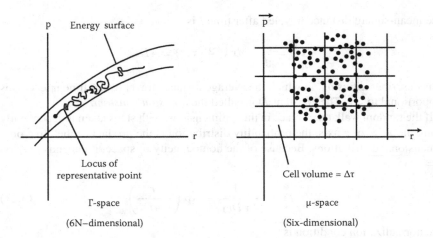

Figure 5.2 In Γ-space, the evolution of an N-particle system is represented by one trajectory on a $(6N - 1)$-dimensional energy surface. In μ-space, it is represented by the "billowing" of a cloud of N points in a six-dimensional space.

given initial conditions. Because of energy conservation, it always lies on an *energy surface*, a hypersurface in Γ-space defined by

$$H(p, r) = E \tag{5.27}$$

Because of atomic collisions, the trajectory is jagged. It can never intersect itself, for the dynamics determines a unique trajectory from a given initial state, and it is time-reversible. Thus the trajectory makes what appears to be a self-avoiding random walk on the energy surface. A symbolic representation of Γ-space is indicated in Figure 5.2.

Another way to specify the state of the system is to describe each atom separately. The motion of each atom is describe by momentum and position $(\mathbf{p}, \mathbf{r})$, which spans a six-dimensional phase space called the μ-space. The overall system is represented by $N \approx 10^{19}$ points, as illustrated schematically in Figure 5.2. These points move and collide with one another as time goes on, and the aggregate billows like a cloud.

In an ideal gas, which is the low-density limit of a real gas, atoms can be thought of hard spheres that do not interact except when they collide. The total energy is well approximated by the sum of energies of the individual atoms. However, collisions cannot be ignored, for they are responsible for the establishment of thermal equilibrium.

5.4 Distribution Function

We are not interested in the behavior of individual atoms, but statistical properties of the entire system, for that is what we can observe on a macroscopic scale. Such statistical properties can be obtained from the distribution function $f(\mathbf{p}, \mathbf{r}, t)$ defined as follows. Divide μ-space into cells, which are six-dimensional volume elements

$$\Delta\tau = \Delta p_x \Delta p_y \Delta p_z \Delta x \Delta y \Delta z \tag{5.28}$$

A cell is assumed to be large enough to contain a large number of atoms, and yet small enough to be considered infinitesimal on a macroscopic scale. From a macroscopic point of view, atoms in the ith cell have unresolved positions $\mathbf{r}_i$, and momenta $\mathbf{p}_i$, and a common energy $\epsilon_i = \mathbf{p}_i^2/2m$.

The number of atoms in cell i at time t is called the *occupation number* n_i. The distribution function is the occupation number per unit volume:

$$f(\mathbf{p}_i, \mathbf{r}_i, t)\Delta\tau = n_i \tag{5.29}$$

Since there are N atoms with total energy E, we have the conditions

$$\sum_i n_i = N$$

$$\sum_i n_i \epsilon_i = E \tag{5.30}$$

The unit for the phase-space volume $\Delta\tau$ is so far arbitrary. This does not lead to ambiguities when $f\Delta\tau$ appear together, but when f appears alone, as in the expression for entropy later, we will have an undetermined constant. As we shall see, quantum mechanics determines the unit to be h^3, where h is Planck's constant.

We assume that $f(\mathbf{p}, \mathbf{r}, t)$ approaches a continuous function in the thermodynamic limit, and regard $\Delta\tau$ as mathematically infinitesimal:

$$\Delta\tau \to d^3p\, d^3r \tag{5.31}$$

We can then write

$$\int d^3p\, d^3r\, f(\mathbf{p}, \mathbf{r}, t) = N$$

$$\int d^3p\, d^3r\, f(\mathbf{p}, \mathbf{r}, t)\frac{\mathbf{p}^2}{2m} = E \tag{5.32}$$

If the density is uniform, then $f(\mathbf{p}, \mathbf{r}, t)$ is independent of $\mathbf{r}$. We denote it by $f(\mathbf{p}, t)$, and write

$$\int d^3p\, f(\mathbf{p}, t) = \frac{N}{V}$$

$$\int d^3p\, f(\mathbf{p}, t)\frac{\mathbf{p}^2}{2m} = \frac{E}{V} \tag{5.33}$$

The distribution function evolves in time according to microscopic equations of motion, and we assume that it eventually approaches a time-independent form $f_0(\mathbf{p}, \mathbf{r})$, which corresponds to thermal equilibrium. Our task is to find the equilibrium distribution, and to deduce from it the thermodynamics of the ideal gas.

5.5 Ergodic Hypothesis

After a time long compared to the collision time, (which is about 10^{-10} s for a gas at STP,) the system should reach some kind of steady state that corresponds to thermal equilibrium.

In classical mechanics, the motion of more than three bodies is generally "chaotic." That is, two states initially close to each other will diverge from each other exponentially with time. Thus, a small change in the initial condition will lead to very different final states, after a long time. This is the basis of the expectation that the trajectory in Γ-space becomes a random walk. This is more precisely stated as the *ergodic hypothesis*:

> Given a sufficiently long time, the representative point of an isolated system will come arbitrarily close to any given point on the energy surface.

The statement can be proven for systems with certain mathematical properties that are somewhat artificial. However, even if one could extend the proof to realistic systems, it does not provide a criterion for "sufficiently long time." Most mathematical attempts to prove the hypothesis use methods that avoid dynamics, whereas the physically relevant question of the relaxation time is a dynamical one. Thus, although the ergodic theorem gives us conceptual understanding, it does not hold any promise for practical applications.

5.6 Statistical Ensemble

In making measurements we effectively perform time averages, because instruments have finite resolutions. The pressure read on a manometer, for example, is a time average over the response time of the instrument, which may be a tiny fraction of a second, but extremely long compared to the collision time. In the statistical treatment, we assume that a time average can be replaced by an average over a suitably chosen collection of systems called a *statistical ensemble*. It is conceptually an infinite collection of identical copies of the system, characterized by a density function $\rho(p, r, t)$ in Γ-space:

$$\rho(p, r, t)dp\, dr = \text{number of system in } dp\, dr \text{ at time } t \qquad (5.34)$$

where (p, r) denotes all the momenta and coordinates of the particles in the system:

$$(p, r) = (\mathbf{p}_1, \dots, \mathbf{p}_N\,;\; \mathbf{r}_1, \dots, \mathbf{r}_N)$$
$$dp\, dr = d^{3N}p\, d^{3N}r \qquad (5.35)$$

The probability per unit phase-space volume of finding the system in $dp\,dr$ at time t is given by

$$\text{Probability density} = \frac{\rho(p,r,t)}{\int dp\,dr\,\rho(p,r,t)} \qquad (5.36)$$

The *ensemble average* of a physical quantity $O(p,r)$ is defined as in the thermodynamic limit

$$\langle O \rangle = \frac{\int dp\,dr\,\rho(p,r,t)\,O(p,r)}{\int dp\,dr\,\rho(p,r,t)} \qquad (5.37)$$

It is important to keep in mind that members of the ensemble are mental copies of the system and do not interact with one another.

As the system approaches thermal equilibrium, the ensemble evolves into an equilibrium ensemble with a time-independent density $\rho(p,r)$. The ensemble average with respect to $\rho(p,r)$ then yields thermodynamic quantities. We assume that $\rho(p,r)$ to depend on (p,r) only through the Hamiltonian, and denote it by $\rho(H(p,r))$. This automatically makes it time-independent, since the Hamiltonian is a constant of the motion.

5.7 Microcanonical Ensemble

For an isolated system, the density ρ is constant over an energy surface, according to the ergodic hypothesis. This condition is known as the assumption of equal *a priori* probability, and defines the *microcanonical ensemble*:

$$\rho(H(p,r)) == \begin{cases} 1 & (\text{if } E < H(p,r) < E + \Delta) \\ 0 & (\text{otherwise}) \end{cases} \qquad (5.38)$$

where Δ is some fixed number that specifies the tolerance of energy measurements, with $\Delta \ll E$. The volume occupied by the microcanonical ensemble is, up to a constant factor specifying the units,

$$\Gamma(E,V) = \int dp\,dr\,\rho\,(H(p,r)) = \int_{E<H(p,r)<E+\Delta} dp\,dr \qquad (5.39)$$

where the dependence on the spatial volume V comes from the limits of the integrations over dr. This is the volume of the shell bounded by the two energy surfaces with respective energies $E + \Delta$ and E. Since $\Delta \ll E$, it can be obtained by multiplying the surface area of the energy surface E by the thickness Δ of the shell. The surface area, in turn, can be obtained from the volume of the interior of the surface. Thus,

$$\Gamma(E,V) = \frac{\partial \Phi(E,V)}{\partial E}\Delta$$

$$\Phi(E,V) = \int_{H(p.q)<E} dp\,dr \qquad (5.40)$$

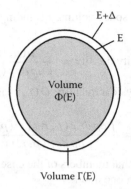

Figure 5.3   The volume Γ of the thin shell can be obtained from the volume Φ of the interior of the shell: $\Gamma = (\partial\Phi/\partial E)\Delta$.

where $\Phi(E, V)$ is the volume of phase space enclosed by the energy surface E, as illustrated in Figure 5.3. Ensemble averages are independent of the unit used for Γ.

The connection to thermodynamics is furnished by the definition of entropy:

$$S(E, V) = k_B \ln \Gamma(E, V) \tag{5.41}$$

The factor k_B, Boltzmann's constant, establishes the unit. The arbitrariness of the units for Γ means that S is defined up to an arbitrary additive constant. The reason for taking the logarithm of Γ is to make S additive for independent systems with the same number of particles. Two independent noninteracting systems have separate distributions ρ and ρ', respectively, and they occupy volumes Γ and Γ' in their respective Γ-spaces. The total Γ-space is the direct product of the two spaces and the total volume is the product $\Gamma\Gamma'$. Another way of stating this is that the probability of two independent events is the product of the individual probabilities. Thus the total entropy is

$$k_B \ln(\Gamma\Gamma') = k_B \ln \Gamma + k_B \ln \Gamma' \tag{5.42}$$

This definition yields $S(E, V)$, the entropy as a function of energy E and volume V, for a fixed number of particles. To obtain thermodynamic functions through the Maxwell relations, however, we need to obtain the function $E(S, V)$ from $S(E, V)$. This procedure is somewhat cumbersome, but we can circumvent it by calculating ensemble averages, as we shall see.

A more convenient ensemble for calculations is the canonical ensemble, which we shall take up in a later chapter.

5.8   Correct Boltzmann Counting

In defining the distribution function, we have discretized the μ-space of a single atom into cells labeled by i, and denote the occupation number of the cell by n_i.

We refer to the set of occupation numbers $\{n_i\}$ as a distribution. In classical physics, this distribution refers to not one microscopic state but a multitude of states. This is because the permutation of the coordinate of two atoms leads to a new microscopic state of the gas, but it does not change the distribution. Therefore a given distribution $\{n_i\}$ corresponds to a collection of states in Γ-space; it "occupies" a volume in Γ-space, which we denote by

$$\Omega\{n_i\} \equiv \Omega\{n_1, n_2, \ldots\} \tag{5.43}$$

The total volume in Γ-space is obtained by adding the contributions from all distributions:

$$\Gamma(E, V) = \sum_{\{n_i\}} \Omega\{n_1, n_2, \ldots\} \tag{5.44}$$

where the sum $\sum_{\{n_k\}}$ extends over all possible sets $\{n_k\}$ that satisfy the constraints [Equation (5.30)]. The ensemble average of the occupation number n_k is given by

$$\langle n_k \rangle = \frac{\sum_{\{n_i\}} n_k \Omega\{n_1, n_2, \ldots\}}{\sum_{\{n_i\}} \Omega\{n_1, n_2, \ldots\}} \tag{5.45}$$

In computing $\Omega\{n_i\}$, there is what appears to be an ad hoc rule, known as "correct Boltzmann counting." Picture the cells of μ-space as a array of boxes with n_i identical balls in the ith box, as illustrated in Figure 5.4. The number of microscopic states corresponding to this distribution is the number of permutations of the N balls that leaves the distribution unchanged:

$$\Omega_0\{n_i\} = \frac{N!}{n_1! n_2! \cdots n_K!} \lambda_1^{n_1} \cdots \lambda_K^{n_K} \tag{5.46}$$

where λ_i is the intrinsic probability for the ith cell. Correct Boltzmann counting instructs us to omit the factor $N!$ and take

$$\Omega\{n_i\} = \frac{1}{n_1! n_2! \cdots n_K!} \lambda_1^{n_1} \cdots \lambda_K^{n_K} \tag{5.47}$$

This rule is rooted in the quantum-mechanical notion of indistinguishability, which has no classical analog. There is no logical way to justify it within classical mechanics itself, but it is necessary in order to make the entropy per particle a function of the density N/V, as is required of a truly extensive quantity.

The intrinsic probabilities λ_i are introduced for mathematical convenience, and will eventually be set to unity.

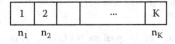

Figure 5.4 Place n_1 balls in box 1, n_2 balls in box 2, and so on.

5.9   Distribution Entropy: Boltzmann's H

The probability of finding an atom in state i is given by n_i/N. If the distribution $\{n_i\}$ is sharply peaked about some i, we are relatively certain where to find an atom. If the distribution is broad, on the other hand, we would be more uncertain about the state of an atom. We may look upon the distribution as a measure of uncertainty, and it is quantified by Boltzmann's H function:[2]

$$H_B\{n_i\} = k_B \ln \Omega\{n_i\} \tag{5.48}$$

which might be called the "distribution entropy".

From Equation (5.47) we have

$$\ln \Omega\{n_i\} = -\sum_i \ln n_i! + \sum_i n_i \ln \lambda_i \tag{5.49}$$

Assuming that n_i are large compared to unity, we can use the Sterling approximation $\ln n! \approx n \ln n - n$ (see Appendix). Thus,

$$H_B\{n_i\} = -k_B \sum_i n_i \ln n_i - k_B \sum_i (\lambda_i - 1)\, n_i \tag{5.50}$$

The last term vanishes when we put $\lambda_i = 1$, as we must do eventually. The form of Boltzmann's H can also be derived in the context of information theory, from consistency requirements relating to dividing sets into subsets, as we shall see later.

The thermodynamic entropy of the system is, according to Equations (5.41) and (5.44),

$$\frac{S}{k_B} = \ln \sum_{\{n_i\}} \Omega\{n_i\} = \ln \sum_{\{n_i\}} \exp \frac{H_B\{n_i\}}{k_B} \tag{5.51}$$

which looks somewhat complicated. In the thermodynamic limit, however, the sum over distributions is dominated by the most probable distribution $\{\bar{n}_i\}$, which maximizes H_B. Thus, for all practical purposes,

$$S = H_B\{\bar{n}_i\} = -k_B \sum_i \bar{n}_i \ln \bar{n}_i \tag{5.52}$$

We shall calculate this explicitly in Section 6.5, and show that it agrees with a calculation using the basic definition (Problem 5.7). The dominance of the most probable distribution will be explained in Section 6.7.

---

[2]Boltzmann's H was meant to be the Greek capital eta. His famous "H theorem" states that the distribution entropy never decreases under the action of atomic collisions. It was the first proof of the second law of thermodynamics from atomic theory. For more details, see Huang 1987.

5.10 The Most Probable Distribution

The *most probable distribution* $\{\bar{n}_i\}$ is the set of occupation numbers that maximizes H_B under the constraints [Equation (5.30)]. In principle, we would like to obtain the ensemble average $\langle n_k \rangle$, but $\bar{n}_k$ is easier to calculate, and we can show that the two coincide in the thermodynamic limit. (See Chapter 6)

To find $\{\bar{n}_i\}$, we note that, since $H_B\{\bar{n}_i\}$ is at a maximum, an infinitesimal change away from $\{\bar{n}_i\}$ will produce a change of second-order smallness. That is, under the variation $n_i \rightarrow \bar{n}_i + \delta n_i$, the entropy change is $\delta H_B = O(\delta n_i)^2$. Therefore the most probable distribution is determined by the condition

$$\delta H_B\{n_i\} = 0 \tag{5.53}$$

with the constraints

$$\sum_i \delta n_i = 0$$

$$\sum_i \epsilon_i \delta n_i = 0 \tag{5.54}$$

Because of these constraints, we cannot vary the n_i independently, but must use the method of Lagrange multipliers (derived in the Appendix). That is, we consider a modified problem

$$\delta \left[\frac{H_B\{n_i\}}{k_B} + \alpha \sum_i n_i + \beta \sum_i \epsilon_i n_i \right] = 0 \tag{5.55}$$

where α and β are fixed parameters called Lagrange multipliers. We can now vary each n_i independently, and after obtaining $\bar{n}_i$ as a function of α and β, we determine α and β so as to satisfy the constraints [Equation (5.30)].

Varying the n_i independently, we obtain

$$\sum_i [-\ln n_i + \alpha + \beta \epsilon_i + \ln \lambda_i] \delta n_i = 0 \tag{5.56}$$

Since the δn_i are arbitrary and independent, we must have

$$\ln n_i = \alpha - \beta \epsilon_i + \ln \lambda_i$$

$$n_i = \lambda_i C e^{-\beta \epsilon_i} \tag{5.57}$$

where $C = e^\alpha$. We now set $\lambda_i = 1$ and write

$$n_i = C e^{-\beta \epsilon_i} \tag{5.58}$$

The Lagrange multipliers are determined by the conditions

$$C \sum_i e^{-\beta \epsilon_i} = N$$

$$C \sum_i \epsilon_i e^{-\beta \epsilon_i} = E \qquad (5.59)$$

In the limit $N \to \infty$, we make the replacement

$$n_i \to f(\mathbf{p}_i, \mathbf{r}_i) d^3 p d^3 r \qquad (5.60)$$

For a free gas in the absence of external potential, the distribution function is independent of $\mathbf{r}$:

$$f(\mathbf{p}) = C e^{-\beta \mathbf{p}^2 / 2m} \qquad (5.61)$$

Thus the conditions become

$$\int d^3 p \, f(\mathbf{p}) = \frac{N}{V} = n$$

$$\frac{1}{n} \int d^3 p \, \frac{\mathbf{p}^2}{2m} f(\mathbf{p}) = \frac{E}{N} \qquad (5.62)$$

The distribution function $f(\mathbf{p})$ is called the *Maxwell–Boltzmann distribution*. It describes a uniform free gas in thermal equilibrium, with density n and energy per particle E/N. We shall study its properties in detail in the next chapter.

5.11   Information Theory: Shannon Entropy

To derive the distribution entropy from a different perspective, we turn to information theory, which is concerned with measuring the uncertainty in a message transmitted through a noisy channel.

A message is made up of characters x, which can assume K possible values $\{x_1, \ldots, x_K\}$ (e.g., $K = 128$ for the characters in ASCII.) The uncertainty in a character x is described by a probability distribution $\{p_1, \ldots, p_K\}$, where p_i is the probability that $x = x_i$. Clearly, $p_i \geq 0$ and $\sum_i p_i = 1$. To map this problem into that of the atom gas, we make the identification $p_i = n_i / N$, and allow K to approach infinity.

Given a probability distribution $\{p_1, \ldots, p_K\}$ how can we quantify the uncertainty inherent in the distribution? (Shannon 1948) proposes the following entropy function

$$H_S\{p_i\} \equiv H_S(p_1, \ldots, p_K) \quad \text{(Shannon entropy)} \qquad (5.63)$$

which is a continuous function that satisfies a certain "composition rule," as explained below.

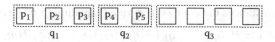

Figure 5.5 A set of cells is grouped into arbitrary boxes of varying sizes. To identify a cell, we can first point to a box, then point to a location inside the box. The defining property of the Shanon entropy, which measures the uncertainty associated with a distribution over the cells, is that it be independent of how the boxes are chosen.

Picture the possible values $\{x_1, \ldots, x_K\}$ as cells illustrated in Figure 5.5. We can group the cells into boxes of various sizes in an arbitrary manner. To point to a cell, we can first point to a box and then point to a location inside the box. The *composition rule* says that the entropy must be independent of how we choose to box the cells.

Let us put the first a cells into box 1, the next b cells into box 2, and so on. The probability associated with a box is the sum of cell probabilities inside the box:

$$q_1 = p_1 + \cdots + p_a$$

$$q_2 = p_{a+1} + \cdots + p_{a+b}$$

$$\cdots \tag{5.64}$$

The conditioned probability of cell i, when it is known that it is contained in box λ, is p_i/q_λ. We can see this is by noting that it should reduce to p_i when multiplied by the probability of box λ. The composition rule states

$$H_S(p_1, \ldots, p_n) = H_S(q_1, \ldots, q_m) + q_1 H_S\left(\frac{p_1}{q_1}, \ldots, \frac{p_a}{q_1}\right)$$

$$+ q_2 H_S\left(\frac{p_{a+1}}{q_2}, \ldots, \frac{p_{a+b}}{q_2}\right) + \cdots \tag{5.65}$$

The first term on the right side refers to boxes; the subsequent terms refer to locations within the boxes. The latter are multiplied by the probability of the boxes, because an additional uncertainty is incurred only when a box is pointed to.

We now find H_S by solving Equation (5.65). It is sufficient to find H_S for rational arguments, since it is a continuous function. For the special case of a uniform distribution, with $p_i = 1/K$, let H_S be denoted by

$$J(K) = H_S\left(\frac{1}{K}, \ldots, \frac{1}{K}\right) \tag{5.66}$$

Now group the cells into B boxes, with m_λ cells in box λ. We must have $\sum_\lambda m_\lambda = K$. The probabilities of box λ is given by

$$q_\lambda = \frac{m_\lambda}{\sum_\beta m_\beta} \tag{5.67}$$

Under these circumstances the composition rule [Equation (5.65)] reduces to

$$J(K) = H_S(q_1, \ldots, q_B) + q_1 J(m_1) + \cdots + q_B J(m_B) \tag{5.68}$$

which can be rewritten as

$$H_S(q_1, \ldots, q_B) = J\left(\sum_{\lambda=1}^{B} m_\lambda\right) - \sum_{\lambda=1}^{B} q_\lambda J(m_\lambda) \tag{5.69}$$

Now choose all $m_\lambda = s$, so that

$$\sum_{\lambda=1}^{B} m_\lambda = Bs$$

$$q_\lambda = \frac{1}{B} \tag{5.70}$$

Then Equation (5.69) reduces to

$$J(B) + J(s) = J(Bs) \tag{5.71}$$

the solution of which is[3]

$$J(z) = C_0 \ln z \tag{5.72}$$

where C_0 is an arbitrary constant. Substituting this into Equation (5.69), we obtain

$$H_S(q_1, \ldots, q_B) = C_0 \ln \left(\sum_{\lambda=1}^{B} m_\lambda\right) - C_0 \sum_{\lambda=1}^{B} q_\lambda \ln(m_\lambda) \tag{5.73}$$

Noting from Equation (5.67) that $m_\lambda = q_\lambda \sum_\beta m_\beta$, we obtain

$$H_S\{q_\lambda\} = -C_0 \sum_\lambda q_\lambda \ln q_\lambda \tag{5.74}$$

which, up to units and an additive constant, is none other than Boltzmann's H [Equation (5.50)].

In information theory, it is customary to put $C_0 = 1$ and replace the natural logarithm ln by $\log_2$, the logarithm to base 2. The unit of the Shannon entropy is then called a "bit."

Problems

5.1 Hydrogen gas is contained in a cylinder at STP. Estimate the number of times the wall of the cylinder is being hit by atoms per second, per unit area.

5.2 A room of volume $3 \times 3 \times 3$ m^3 contains air at STP. Treating the air molecules as independent objects, estimate the probability that you will find a 1-cm^3 volume

---

[3] Shannon (1948) shows that the solution is unique.

somewhere in the room totally devoid of air, due to statistical fluctuations. Do the same for a 1-A^3 volume.

5.3 In a gas at STP, let $p(r)dr$ be the probability that an atom has a nearest neighbor between distances r and $r + dr$. Find $p(r)$.

5.4 In an atomic beam experiment a collimated beam of neutral Na atoms traverses a vacuum chamber for a distance of 10 cm. How good a vacuum is required for the beam to remain well-collimated during the transit?

5.5 Neutrinos are the most weakly interacting particles we know of, and they can penetrate the earth with the greatest of ease. This has caused some writer to worry about neutrinos hitting "a lover and his lass" from beneath the bed (by way of Nepal). Is the fear founded?

(a) Assuming a neutrino cross section of $\sigma = 10^{-40}$ cm^2 for collision with a nucleon, estimate the neutrino mean free path in water.

(b) Sources for neutrinos include cosmic-ray reactions in the atmosphere, and astrophysical events. For the sake of argument, assume there is a neutrino flux of 50 cm^{-2} s^{-1}. Show that a person whose mass is of order 100 kg would get hit by a neutrino about once in a lifetime. (Perhaps that's what kills the person.)

5.6 Volume of n-sphere
To calculate the phase-space volume of an ideal gas in the microcanonical ensemble, we need the surface area of an n-dimensional sphere of radius R. The volume of an n-sphere of radius R is of the form

$$\Phi_n(R) = C_n R^n$$

The surface area is

$$\Sigma_n(R) = nC_n R^{n-1}$$

All we need is the constant C_n. Show that

$$C_n = \frac{2\pi^{n/2}}{\Gamma\left(\frac{n}{2} + 1\right)}$$

where $\Gamma(z)$ is the gamma function.
Suggestion: We know that $\int_{-\infty}^{\infty} dx \exp(-\lambda x^2) = \sqrt{\pi/\lambda}$. Thus

$$\int_{-\infty}^{\infty} dx_1 \cdots dx_n \, e^{-\lambda(x_1^2 + \cdots + x_n^2)} = (\pi/\lambda)^{n/2}$$

Now rewrite the integral in spherical coordinates as $\int_0^{\infty} dR \, R^{n-1} \exp(-\lambda R^2)$, which is a gamma function.

5.7 Entropy of ideal gas
The Hamiltonian for a free classical ideal gas of N atoms can be written as $H = p_1^2 + p_2^2 + \cdots + p_{3N}^2$, where we have chosen units such that $2m = 1$.

(a) Show that the phase-space volume is $\Gamma(E, V) = K_0 V^N \Sigma_n(E)$, where K_0 is a constant, and $n = 3N$.

(b) Calculate Σ_n, and show that the entropy $S(E, V)$ in the thermodynamic limit agrees with the result from thermodynamic:

$$\frac{S(E, V)}{Nk_B} = \ln(V E^{3/2}) + O\left(\frac{\ln N}{N}\right)$$

5.8 Ideal gas in harmonic trap

If we put the ideal gas in an external harmonic-oscillator trapping potential, the Hamiltonian would become $H = (p_1^2 + p_2^2 + \cdots + p_{3N}^2) + (r_1^2 + r_2^2 + \cdots + r_{3N}^2)$, in special units.

(a) Show that the phase-space volume is $\Gamma(E, V) = K_1 \sigma_n(E)$, where K_1 is a constant, and $n = 6N$.

(b) Find the entropy of the system.

5.9 Model the trajectory of a molecule in a gas as a random walk in 3D, due to collisions. Give an order-of-magnitude estimate of the time it would take an air molecule in a room to traverse a distance of 1 cm. What about 1 m?

5.10 In the free-expansion experiment illustrated in Figure 2.1, a gas of N atoms was originally confined inside a compartment. A hole was opened to let the gas expand into the next compartment, which was originally vacuous. Suppose each gas atom executes random walk due to collisions. In principle, all the atoms can get back to the initial state by sheer coincidence. Show that, for that to happen, one would have to wait the order of e^N collision times. How many ages of the universe does this correspond to?

References

Huang, K., *Statistical Mechanics*, 2nd ed., Wiley, New York, 1987, Chapter 4.

Shannon, C.E., *Bell System Tech. J.*, **27**:379, 623 (1948).

Shannon, C.E. and W. Weaver, *The Mathematical Theory of Communications*, University of Illinois Press, Urbana, 1949.

Chapter 6

Maxwell–Boltzmann Distribution

6.1 Determining the Parameters

The Maxwell–Boltzmann distribution for an ideal gas is

$$f(\mathbf{p}) = Ce^{-\beta p^2/2m} \tag{6.1}$$

The conditions that determine the parameters C and β are

$$\int d^3p\, f(\mathbf{p}) = n$$

$$\frac{1}{n} \int d^3p\, \frac{p^2}{2m} f(\mathbf{p}) = \frac{E}{N} \tag{6.2}$$

where n is the density of the gas and E/N is the energy per particle.

To explicitly calculate C and β, we need the Gaussian integral

$$\int_{-\infty}^{\infty} dx\, e^{-\lambda x^2} = \sqrt{\frac{\pi}{\lambda}} \tag{6.3}$$

Related integrals can be obtained by differentiating both sides with respect to λ:

$$\int_{-\infty}^{\infty} dx\, x^2\, e^{-\lambda x^2} = \frac{\sqrt{\pi}}{2\lambda^{3/2}}$$

$$\int_{-\infty}^{\infty} dx\, x^4 e^{-\lambda x^2} = \frac{3\sqrt{\pi}}{4\lambda^{5/2}} \tag{6.4}$$

Thus

$$n = C \int_{-\infty}^{\infty} dp_1 dp_2 dp_3\, e^{-\lambda(p_1^2 + p_2^2 + p_3^2)}$$

$$= C \left[\int_{-\infty}^{\infty} dp\, e^{-\lambda p^2} \right]^3 = C \left(\frac{\pi}{\lambda} \right)^{3/2} \tag{6.5}$$

and

$$C = n \left(\frac{\lambda}{\pi} \right)^{3/2} \tag{6.6}$$

Putting $\lambda = \beta/(2m)$, we obtain

$$C = n \left(\frac{\beta}{2\pi m} \right)^{3/2}. \tag{6.7}$$

Next we calculate

$$
\begin{aligned}
\frac{E}{N} &= \frac{C}{2mn} \int_{-\infty}^{\infty} dp_1 dp_2 dp_3 \left(p_1^2 + p_2^2 + p_3^2 \right) e^{-\lambda(p_1^2+p_2^2+p_3^2)} \\
&= \frac{3}{2m} \left(\frac{\lambda}{\pi} \right)^{3/2} \int_{-\infty}^{\infty} dp_1 dp_2 dp_3 \, p_1^2 \, e^{-\lambda(p_1^2+p_2^2+p_3^2)} \\
&= \frac{3}{2m} \left(\frac{\lambda}{\pi} \right)^{3/2} \left[\int_{-\infty}^{\infty} dp_1 \, p_1^2 \, e^{-\lambda p_1^2} \right] \left[\int_{-\infty}^{\infty} dp \, e^{-\lambda p^2} \right]^2 \\
&= \frac{3}{2m} \left(\frac{\lambda}{\pi} \right)^{3/2} \frac{\sqrt{\pi}}{2\lambda^{3/2}} \left(\frac{\pi}{\lambda} \right) = \frac{3}{4m\lambda} = \frac{3}{2\beta} \tag{6.8}
\end{aligned}
$$

Therefore

$$\beta = \frac{3}{2} \frac{E}{N} \tag{6.9}$$

We shall now show that $\beta = (k_B T)^{-1}$.

6.2   Pressure of Ideal Gas

The pressure of an ideal gas is the average force per unit area that it exerts on the wall of its container. Take the wall to be normal to the x axis and assume that the wall is perfectly reflecting. When an atom with x-component velocity v_x is reflected by the wall, it will transfer an amount of momentum $2mv_x$ to the wall. The force acting on the wall is the momentum transfer per unit time and the pressure is the force per unit area N the wall:

$$\text{Pressure} = (\text{momentum transfer per atom}) \times (\text{flux of atoms}) \tag{6.10}$$

The flux is the number of atoms crossing unit area per second, and this is equal to the number of atoms contained in a cylinder of length equal to v_x, of unit cross section. This is illustrated in Figure 6.1. Thus,

$$\text{Flux of atoms} = v_x \, f(p) \, d^3 p \tag{6.11}$$

The pressure is given by

$$P = \int_{v_x>0} d^3 p (2mv_x) v_x \, f(\mathbf{p}) = m \int d^3 p \, v_x^2 \, f(\mathbf{p}) \tag{6.12}$$

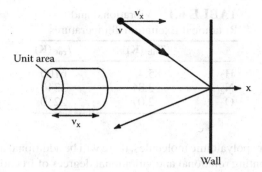

Figure 6.1 In 1 second, all the atoms with x-component velocity v_x would have evacuated a cylinder of length numerically equal to v_x.

In the integrand, we may replace v_x^2 by $\frac{1}{3}(v_x^2 + v_y^2 + v_z^2) = \mathbf{p}^2/3m^2$. Thus,

$$P = \frac{1}{3m} \int d^3 p\, \mathbf{p}^2\, f(\mathbf{p}) = \frac{2}{3} \frac{E}{V} \qquad (6.13)$$

The ideal gas temperature T, which coincides with the absolute temperature, is defined by $PV = Nk_B T$. Thus

$$\frac{E}{N} = \frac{3}{2} k_B T \qquad (6.14)$$

Since $E/N = 3/(2\beta)$, as shown earlier, we have $\beta = (k_B T)^{-1}$.

In summary, the parameters in the Maxwell–Boltzmann distribution are given by

$$\beta = \frac{1}{k_B T}$$

$$C = \frac{n}{(2\pi m k_B T)^{3/2}} \qquad (6.15)$$

6.3 Equipartition of Energy

The factor 3 in the formula $E/N = \frac{3}{2} k_B T$ represents the number of translational degrees of freedom, the number of momentum components appearing in the energy of an atom:

$$\epsilon = \frac{1}{2m} \left(p_x^2 + p_y^2 + p_z^2 \right) \qquad (6.16)$$

Each quadratic term in the energy contributes $k_B T/2$ to the internal energy per particle, hence $k_B/2$ to the specific heat at constant volume. This is known as the *principle of equipartition of energy*.

TABLE 6.1 Vibrational and
Rotational Excitation Temperatures

	T_{vib} (K)	T_{rot} (K)
H_2	85.4	6100
N_2	2.86	3340
O_2	2.07	2230

If our "atoms" are polyatomic molecules, there will be additional quadratic terms in the energy, representing rotational and vibrational degrees of freedom. Each of these term will contribute $k_B/2$ to the specific heat additively, as long as the thermal energy $k_B T$ is sufficient to excite them quantum-mechanically. For example, a diatomic molecule such as H_2 has two rotational and one vibrational degrees of freedom. Its energy is of the form

$$\epsilon = \frac{1}{2M} \left(P_x^2 + P_y^2 + P_z^2 \right) + \frac{J_\perp^2}{2I_\perp} + \frac{J_\parallel^2}{2I_\parallel} + \left(\frac{p^2}{2\mu} + \frac{\mu\omega_0^2}{2}q^2 \right) \qquad (6.17)$$

where P is the total momentum, and $J_\perp$ and $J_\parallel$, respectively, denote the angular momentum perpendicular and parallel to the symmetry axis of the molecule. The last two terms represent the energy of a harmonic oscillator corresponding to the vibrational mode. These quantities have to be treated as quantum operators. In particular, the rotational and vibrational energies have discrete eigenvalues, and have minimal energies $k_B T_{rot}$, and $k_B T_{vib}$, respectively. Examples of these threshold values are listed in Table 6.1 (data from Wilson 1957).

At room temperature the thermal energy is not sufficient to excite the vibrational mode of H_2, and so its specific heat is $\frac{5}{2}k$. Figure 6.2 shows the heat capacity of 1 mol

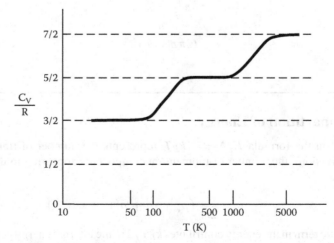

Figure 6.2 Heat capacity per mole of H_2 as function of temperature in log scale. (After Present 1958.)

of molecular hydrogen over a wide range of temperatures on a logarithmic scale. Thus we see quantum mechanics at work even at high temperatures.

6.4 Distribution of Speed

The distribution function is independent of position for a gas in the absence of external potentials. This means that atoms move with the same Maxwellian velocity distribution in every volume element in the air around you. Because of the isotropy of space, the distribution depends only on the magnitude p of the momentum. Thus, the components of the velocity average to zero:

$$\langle \mathbf{v} \rangle = \frac{\int d^3 p \, \mathbf{p} \, f(p)}{m \int d^3 p \, f(p)} = 0 \tag{6.18}$$

The mean-square velocity is not zero:

$$\langle \mathbf{v}^2 \rangle = \frac{\langle \mathbf{p}^2 \rangle}{m^2} = \frac{\int d^3 p \, \mathbf{p}^2 f(\mathbf{p})}{m^2 \int d^3 p \, f(\mathbf{p})} = \frac{3 k_B T}{m} \tag{6.19}$$

and leads to a root-mean-square velocity

$$v_{\mathrm{rms}} = \sqrt{\frac{3 k_B T}{m}} \tag{6.20}$$

From this we obtain the mean kinetic energy

$$\frac{1}{2} m \langle \mathbf{v}^2 \rangle = \frac{3}{2} k_B T \tag{6.21}$$

which agrees with the equipartition principle.

The effective volume element in momentum space is $4\pi p^2 dp$. The quantity $4\pi p^2 f(p)$ is the distribution of speed, the number of atoms per unit volume per unit interval of p whose magnitude of momentum lies between p and $p + dp$. A qualitative graph of the speed distribution is shown in Figure 6.3. The area under the entire curve is the particle density n. The area under the curve for $p > p_1$ is the density of particle with magnitude of momenta greater than p_1. The momentum at the maximum is $p_0 = \sqrt{2 m k_B T}$, and the corresponding velocity is called the *most probable velocity:*

$$v_0 = \sqrt{\frac{2 k_B T}{m}} \tag{6.22}$$

If the gas moves as a whole with uniform velocity $\mathbf{v}_0$, then in maximizing $\Omega\{n_i\}$ we must add the constraint that the average momentum per particle is $\mathbf{p}_0 = m \mathbf{v}_0$. This leads to a momentum distribution centered about $\mathbf{p}_0$:

$$f(\mathbf{p}) = C e^{-\lambda(\mathbf{p} - \mathbf{p}_0)^2} \tag{6.23}$$

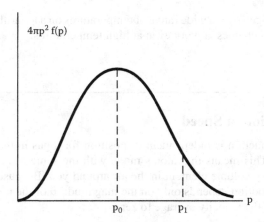

Figure 6.3   Maxwell–Boltzmann distribution of speed.

6.5 Entropy

The entropy, defined as $S = k_B \ln \Gamma$ in Equation (5.41), is calculated in Problem 5.7. We now calculate it according to Equation (5.52), as the distribution entropy of the most probable distribution:

$$S(V, T) = -k_B \sum_i \bar{n}_i \ln \bar{n}_i \qquad (6.24)$$

Writing $\bar{n}_i = f_i \Delta\tau$, where f_i is the Maxwell–Boltzmann distribution function and $\Delta\tau$ the phase space volume element, we have

$$\sum_i \bar{n}_i \ln \bar{n}_i = \Delta\tau \sum_i f_i \ln f_i + N \ln \Delta\tau \qquad (6.25)$$

The last term is an additive constant, and will be ignored. Thus, in the thermodynamic limit we have

$$S(V, T) = -V k_B \int d^3p \, f(\mathbf{p}) \ln f(\mathbf{p}) \qquad (6.26)$$

Now we calculate the integral:

$$\int d^3p \, f(\mathbf{p}) \ln f(\mathbf{p}) = C \int d^3p \, e^{-\lambda \mathbf{p}^2} \ln\left(C e^{-\lambda \mathbf{p}^2}\right)$$

$$= C \int d^3p \, e^{-\lambda \mathbf{p}^2} (\ln C - \lambda \mathbf{p}^2)$$

$$= (C \ln C) \int d^3p \, e^{-\lambda \mathbf{p}^2} - \lambda C \int d^3p \, \mathbf{p}^2 e^{-\lambda \mathbf{p}^2} \qquad (6.27)$$

where $\lambda = (2mk_BT)^{-1}$. In the first term, we note that $C \int d^3p\, e^{-\lambda p^2} = n$. For the second term, explicit calculation gives

$$\lambda C \int d^3p\, \mathbf{p}^2 e^{-\lambda \mathbf{p}^2} = 4\pi \lambda C \int_0^\infty dp p^4 e^{-\lambda \mathbf{p}^2} = \frac{3}{2}n \tag{6.28}$$

Thus

$$\int d^3p\, f(\mathbf{p}) \ln f(\mathbf{p}) = n \ln C - \frac{3}{2}n \tag{6.29}$$

Finally

$$S(V, T) = Nk_B \ln(n^{-1}T^{3/2}) + C_0 \tag{6.30}$$

where C_0 is a constant, which may depend on N. It will be determined in Section 8.7, where we supply correct units for the phase volume. This result agrees with Equation (2.49) obtained in thermodynamics.

The distribution entropy obtained here agrees with the entropy calculated from the basic definition $S = k_B \ln \Gamma$, which includes contributions from all possible distributions. The reason for the agreement lies in the fact that, in the thermodynamic limit, most distributions are close to the most probable one. We shall show this later when we examine fluctuations about the most probable distribution.

6.6 Derivation of Thermodynamics

In the mean time, let us derive the thermodynamics of an ideal gas. The equation of state $PV = Nk_BT$ was obtained earlier. The internal energy is just the total energy of the system:

$$U(T) = E = \frac{3}{2}Nk_BT \tag{6.31}$$

Thus $C_V = \frac{3}{2}Nk$, and $PV = \frac{2}{3}U$. We have calculated the entropy, which can be written in the form

$$\frac{S}{k_B} = N \ln V + \frac{3}{2}N \ln T \tag{6.32}$$

Both U and S are determined only up to an arbitrary additive constant.

Taking the differential of S, we have

$$\frac{dS}{Nk_B} = \frac{dV}{V} + \frac{3}{2}\frac{dU}{U} \tag{6.33}$$

Using the equation of state to write $dV/V = (P/T)dV$, we can rewrite the above as

$$dS = \frac{1}{T}(dU + PdV) \tag{6.34}$$

The first law of thermodynamics is a definition of the heat absorbed:

$$dQ = dU + PdV \qquad (6.35)$$

The second law is the statement

$$dQ = TdS \qquad (6.36)$$

which says T is the integrating factor that makes dQ/T an exact differential.

That the entropy of an isolated system never decreases is implied by the fact that it is a monotonically increasing function of the volume. For an isolated system, the only thing that can happen is that the volume increases, as when a wall of the container of the gas is suddenly withdrawn.

6.7   Fluctuations

We now address the question, "How probable is the most probable distribution?" For an answer, we calculate the mean-square fluctuation about the average occupation $\langle n_i \rangle$. We will show that the fluctuation vanishes when $N \to \infty$, thereby showing that almost all microscopic states of the gas have this distribution, and therefore $\langle n_i \rangle$ coincides with most probable distribution $\bar{n}_i$.

We start with the expression

$$\Omega\{n_i\} = \frac{1}{n_1! n_2! \cdots n_K!} \lambda_1^{n_1} \cdots \lambda_K^{n_K} \qquad (6.37)$$

and take advantage of the presence of the factors λ_k. Taking the partial derivative with respect to λ_k, we have

$$\lambda_k \frac{\partial}{\partial \lambda_k} \Omega\{n_i\} = n_k \Omega\{n_i\} \qquad (6.38)$$

Thus the ensemble average of the occupation number is given by

$$\langle n_k \rangle = \frac{1}{\sum_{\{n_i\}} \Omega\{n_i\}} \lambda_k \frac{\partial}{\partial \lambda_{ik}} \sum_{\{n_i\}} \Omega\{n_i\}$$

$$= \lambda_k \frac{\partial}{\partial \lambda_k} \ln \sum_{\{n_i\}} \Omega\{n_i\} = \frac{\partial \ln \Gamma}{\partial v_k} \qquad (6.39)$$

where

$$v_i = \ln \lambda_i \qquad (6.40)$$

The ensemble average of n_k^2 can be calculated as follows:

$$\langle n_k^2 \rangle = \frac{1}{\sum_{\{n_i\}} \Omega\{n_i\}} \sum_{\{n_i\}} n_k^2 \Omega\{n_i\} = \frac{1}{\Gamma} \frac{\partial^2 \Gamma}{\partial v_k^2}$$

$$= \frac{\partial}{\partial v_k} \left(\frac{1}{\Gamma} \frac{\partial \Gamma}{\partial v_k} \right) + \frac{1}{\Gamma^2} \left(\frac{\partial \Gamma}{\partial v_k} \right)^2$$

$$= \frac{\partial}{\partial v_k} \left(\frac{\partial \ln \Gamma}{\partial v_k} \right) + \left(\frac{\partial \ln \Gamma}{\partial v_k} \right)^2$$

$$= \frac{\partial}{\partial v_k} \langle n_k \rangle + \langle n_k \rangle^2 \qquad (6.41)$$

Thus the mean-square fluctuation is given by

$$\langle n_k^2 \rangle - \langle n_k \rangle^2 = \lambda_k \frac{\partial \langle n_k \rangle}{\partial g_k} \qquad (6.42)$$

Assuming for the moment that

$$\langle n_k \rangle \approx \bar{n}_k = \lambda_k C e^{-\beta \epsilon_{ik}} \qquad (6.43)$$

we have

$$\lambda_k \frac{\partial \langle n_k \rangle}{\partial \lambda_k} = \lambda_k C e^{-\beta \epsilon k} = \langle n_k \rangle \qquad (6.44)$$

Thus

$$\langle n_k^2 \rangle - \langle n_k \rangle^2 = \langle n_k \rangle \qquad (6.45)$$

The fractional fluctuation is given by

$$\frac{\langle n_k^2 \rangle - \langle n_k \rangle^2}{N^2} = \left\langle \frac{n_k}{N} \right\rangle \frac{1}{N} \qquad (6.46)$$

which vanishes like N^{-1} when $N \to \infty$. In the sense we have $\langle n_i \rangle \approx \bar{n}_i$.

We conclude that the Maxwell–Boltzmann distribution is a most prevalent condition for a gas in equilibrium. Imagine that all possible states of the gas with given N and E are placed in a jar. Wearing a blindfold, you pick a state from the jar, and you can expect to get a state with Maxwell–Boltzmann distribution, with overwhelming probability.

6.8 The Boltzmann Factor

Our derivation of the most probable distribution does not specifically assume that we are dealing with a classical gas. It only assumes that we have a collection of noninteracting units with energy ϵ_i for the state i. Thus, we have actually proven a more general result:

> If a system has possible states labelled by i, and the energy of the state i is ϵ_i, then the relative probability for finding the system in state i is given by the Boltzmann factor $e^{-\epsilon_i/k_B T}$. The absolute probability for the occurrence of the state i is
>
> $$\frac{1}{\sum_i e^{-\epsilon_i/k_B T}} e^{-\epsilon_i/k_B T} \qquad (6.47)$$

It should be emphasized that $e^{-\epsilon_i/k_B T}$ is the relative probability for the occurrence of the *state i*, not the energy value ϵ_i. The distinction is important, because different states can have the same energy and this is called a degeneracy in quantum mechanics.

6.9   Time's Arrow

According to the second law of thermodynamics, the entropy of an isolate system can never decrease. This seems to be a valid conclusion, for most events on the macroscopic scale are irreversible. As we all know, it is useless to cry over spilled milk. Therefore, there appears to be an "arrow of time" that points toward an increase of entropy, and distinguishes past from future. How is this to be reconciled with the time-reversal invariance of the microscopic laws of physics?

Consider Figure 6.4, which shows two successive frames of a movie of a gas contained in a partitioned box, with a hole through which the particles can pass. Common sense tells us that frame (a) must precede frame (b), and this establishes the arrow of time.

The equations of motion have solutions for which (a) evolves into (b) or vice versa. In fact we can always reverse the history by reversing all the velocities of the particles instantaneously. However, the situations (a) and (b) are not symmetrical. As an initial state, almost any state that looks like (a) will evolve into a uniform state like (b).

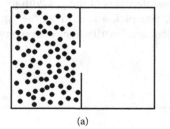 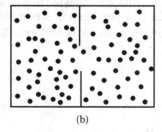

(a)   (b)

Figure 6.4   If these are two successive frames of a movie of particles in a container with partition, common sense tells us that (a) must precede (b), and this establishes "time's arrow."

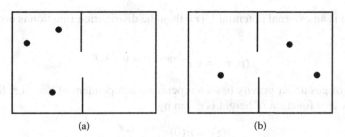

(a)   (b)

Figure 6.5 In this case, it is not so clear which frame precedes which.

But an initial state (b) will not evolve into (a), unless it is very carefully prepared to achieve that purpose.

The point is that, in the $3N$-dimensional phase space of the system, the initial conditions that will make (b) evolve into (a) has a negligible measure, when N is large. On the other hand, most nonuniform states like (a) will develop into a uniform state like (b), because such evolutions are overwhelmingly favored by phase space.

If we start with an "average" initial condition, (b) has a chance to evolve into (a) only after the order of e^N collision times (see Problem 5.11). For $N \approx 100$ as depicted in Figure 6.4, this time is much longer than the age of the universe, which is a mere 10^{17} seconds.

If N is small, as illustrated in Figure 6.5, then the system can evolve from (a) to (b) or vice versa with equal probability, and there is complete reversibility. Time's arrow, therefore, is not inherent in the equation of motion, but is the property of a large-N system set by the initial condition. Spilled milk is irreversible, because someone had prepared the milk in the extremely improbable state of being inside the milk bottle.

In the case of the whole universe, unless the Big Bang was "prepared" in any special way, we must conclude that time's arrow signifies the spontaneous breaking of time-reversal invariance. On the other hand, it is believed that the interaction of elementary particles contains an extremely weak component that violates time-reversal invariance. It is possible that this could have tilted the Big Bang along a preferred time axis.

Problems

6.1 The energy of individual atoms in a gas fluctuates about an average value $\frac{3}{2}k_B T$ because of collisions.

(a) Verify this by calculating $\bar{\epsilon}$, the mean of $\epsilon = p^2/2m$ with respect to the Maxwell–Boltzmann distribution.

(b) Show $\overline{\epsilon^2} - \bar{\epsilon}^2 = \frac{3}{2}(k_B T)^2$.

6.2 Find the energy distribution function $P(E)$ for a classical nonrelativistic ideal gas, such that $P(E)dE$ is the density of atoms with energy between E and $E + dE$.

6.3 If there is an external potential $U(\mathbf{r})$, then the distribution function is nonuniform in space:

$$f(\mathbf{p}, \mathbf{r}) = Ce^{-[(\mathbf{p}^2/2m)+U(\mathbf{r})]/k_BT}$$

A column of gas under gravity has a temperature independent of height z. Show that the density as a function of height is given by

$$n(z) = n(0)e^{-mgz/k_BT}$$

where g is the acceleration of gravity.

6.4 In the atmosphere the temperature varies with height. Assume that there is a steady-state adiabatic convection, that is, no heat transfer in the vertical direction.
   (a) Show that the temperature $T(z)$ changes with height z according to

$$k\frac{dT}{dz} = -\frac{\gamma - 1}{\gamma}mg \qquad (6.48)$$

   (b) Find $T(z)$, and the altitude of the top of the atmosphere where the temperature becomes zero.
   (c) Show that the pressure $P(z)$ changes with height according to

$$\frac{dP}{P} = -\frac{mg}{k_BT(z)}dz$$

Integrate this to find $P(z)$.

6.5 A gas in equilibrium has a distribution function

$$f(\mathbf{p}, \mathbf{r}) = \frac{1 + \gamma x}{k_BT}(2\pi mk_BT)^{-3/2}\exp(-\mathbf{p}^2/2mk_BT)$$

where x is the distance along an axis with a fixed origin, γ is a constant. What is the nature of the force acting on the gas?

6.6 Molecules in a centrifuge rotate about an axis at constant angular velocity ω. In the rotating frame, they are at rest, but experience a centrifugal force $m\omega^2r$, where r is the normal distance from the axis. This is equivalent to an external potential

$$U(r) = -\frac{1}{2}m\omega^2r^2$$

Two dilute gases, of molecular masses m_1 and m_2, respectively, are placed in a centrifuge rotating at a circular frequency ω. Derive the ratio n_1/n_2 of their densities as a function of the distance r from the axis of rotation.

6.7 The Maxwell–Boltzmann distribution for a relativistic gas is

$$f(\mathbf{p}) = Ce^{-\sqrt{p^2+m^2}/k_BT}$$

where we use units in which the velocity of light is $c = 1$.

(a) Find the most probable velocity. Obtain its nonrelativistic ($k_B T \ll m$), and ultrarelativistic ($k_B T \gg m$) limits, both with first-order corrections.

(b) Set up an expression for the pressure. Show that $PV = U/3$ in the ultrarelativistic limit, where U is the average energy.

(c) Find the velocity distribution function $f(\mathbf{v})$, such that $f(\mathbf{v})d^3v$ is the density of particles whose velocity lies in the volume element d^3v. Find the nonrelativistic limit to first order in v/c.

(d) At what temperatures would relativistic effects be important for a gas of H_2 molecules?

6.8 The Doppler formula for the observed frequency f from a source moving with velocity v_x along the line of sight of an observer is

$$f = f_0 \left(1 + \frac{v_x}{c}\right)$$

where f_0 is the frequency in the rest frame of the source.

(a) What is the distribution in frequency of a particular spectral line radiated from a gas at temperature T?

(b) Find the breadth of the line, defined as the variance $\overline{(f - f_0)^2}$.

(c) Atomic hydrogen and atomic oxygen are both present in a hot gas. How much broader is the hydrogen line compared to the oxygen line, of roughly the same frequency?

6.9 Neutrinos are particles whose energy-momentum relation is given by $\epsilon(p) = cp$, where c is the velocity of light. Consider N neutrinos in a volume V, at a temperature T sufficiently high that the system can be treated as a classical gas.

(a) Find the heat capacity C_V of the system.

(b) Find the pressure of the system in terms of the internal energy U. Give it in terms of the temperature.

6.10 If we integrate the Maxwell–Boltzmann distribution over from some momentum up, we are faced with error functions

$$\mathrm{erfc}(y) = \frac{2}{\sqrt{\pi}} \int_y^\infty dx$$

(See Abramowitz and Stegun 1964.)

(a) Show the following asymptotic behavior for large y:

$$\int_y^\infty dx\, e^{-x^2} \approx e^{-y^2} \left[\frac{1}{2y} - \frac{1}{4y^3} + \frac{3}{8}\frac{1}{y^5} + \cdots\right]$$

Hint: Transform to new integration variable $t = x^2$. The asymptotic behavior of the integral $\int_{y^2}^\infty dt\, t^{-1/2} e^{-t}$ can be obtained by repeated partial integrations.

(b) Differentiate $\int_y^\infty dx\, e^{-\lambda x^2}$ with respect to λ, and then set $\lambda = 1$, to obtain the asymptotic formulas

$$\int_y^\infty dx\, x^2 e^{-x^2} \approx \frac{e^{-y^2}}{2y}(y^2 + 1/2)$$

$$\int_y^\infty dx\, x^4\, e^{-x^2} \approx \frac{e^{-y^2}}{2y}(y^4 + y^2 + 3/4)$$

6.11 Suppose a surface of the container of a gas absorbs all molecules striking it with a normal velocity greater than v_0. Find the absorption rate W per unit area.

6.12 The atmosphere contains molecules with high velocities that can escape the Earth's gravitational field.

(a) What fraction of the H_2 gas at sea level, at temperature 300 K, can escape from the Earth's gravitational field?

(b) Give an order-of-magnitude estimate of the time needed for the escape, on the basis of a random walk.

6.13 A gas of N atoms was initially in equilibrium in a volume V at temperature T. In an evaporation process, all atoms with energy greater than $\epsilon_0 = p_0^2/2m$ were allowed to escape, and the gas eventually reestablishes a new equilibrium. Assume $y \equiv \epsilon_0/k_B T \gg 1$.

(a) Find the change ΔN in the number of atoms, and the change ΔE in the energy of the gas, as functions of ϵ_0.

(b) Find the fractional change in temperature $\Delta T/T$ as a function of $\Delta N/N$.

Hint: Find ΔT via $E/N = \frac{3}{2}k_B T$. Express ϵ_0 in terms of ΔN using an iterative process assuming the smallness of the latter.

6.14 The following exercises illustrate the equipartition of energy.

(a) A long thin needle floats in a gas at constant temperature. On the average, is its angular momentum vector nearly parallel to or perpendicular to the long axis of the grain? Explain.

(b) A capacitor $C = 100\ \mu F$ in a passive circuit (no driving voltage) is at temperature $T = 300$ K. Calculate the rms voltage fluctuation.

6.15 An insulated space ship is a cylinder of length L and cross section A filled with air (treated as N_2 gas) at STP. It was brought to a sudden stop from an initial velocity of 7 km/s.

(a) Assuming that a good fraction of the original translational kinetic energy of the air was converted to heat, estimate the temperature rise inside the space ship. Would the astronauts be fried?

(b) Suppose the space ship was stopped by constant deceleration a parallel to the axis of the cylinder, such that the air was in local equilibrium. Show that the pressure difference between the front and back of the space ship is given by

$$\Delta P = P_0 \left(1 - e^{-maL/k_B T}\right)$$

where m is the mass of an air molecule.

(c) Assuming $maL/k_B T \ll 1$, find the force F exerting on the space ship by the air inside, the total work W done by the air, and the temperature rise inside the space ship.

(d) What condition must be imposed on the deceleration to allow the establishment of local equilibrium?

References

Abramowitz, M. and I.A. Stegun (eds.), *Handbook of Mathematical Functions*, National Bureau of Standards, 1964.

Present, R.D., *Kinetic Theory of Gases*, McGraw-Hill, New York, 1958.

Wilson, A.H., *Thermodynamics and Statistical Mechanics*, Cambridge University Press, Cambridge, 1957.

Chapter 7

Transport Phenomena

7.1 Collisionless and Hydrodynamic Regimes

A gas tends toward thermal equilibrium through atomic collisions, which transport mass, momentum, and energy from one part of the system to another. In a collision, these quantities are transported over a mean free path λ, on average. An important parameter in the description of transport phenomena is the ratio λ/L, where L is an external length scale, such as the size of the container, or the wavelength of a density variation. Two extreme limits are amenable to analytical treatment:

$$\text{Collisionless regime:} \quad \lambda \gg L$$

$$\text{Hydrodynamic regime:} \quad \lambda \ll L$$

We can illustrate the two cases by looking at how a gas flows through a hole of dimension L in a wall, as shown in Figure 7.1.

When $\lambda \gg L$, the atom that went through the hole came from a last collision very far from the hole, and will not collide with another atom again until it gets very far from the hole. The passage through the hole can be described by ignoring collisions. A practical example is air leaking into a vacuum system through a very small crack.

When $\lambda \ll L$, an atom makes many collisions during the passage, and thermalizes with the local atoms during its journey. As a consequence, it moves as part of a collective flow. This regime is described by hydrodynamics.

In the collisionless regime, atoms escape through a hole in the wall through *effusion*. Let us set up the x-axis normal to the area of the hole. The flux of atoms through the hole, due to those with momenta lying in the element d^3p, is given by

$$dI = v_x f(\mathbf{p}) d^3 p \tag{7.1}$$

The total flux is then

$$I = \int_{v_x > 0} v_x f(\mathbf{p}) d^3 p$$

$$= C \int_0^\infty dp_x \frac{p_x}{m} e^{-\lambda p_x^2} \left[\int_{-\infty}^\infty dp_y e^{-\lambda p_y^2} \right]^2 \tag{7.2}$$

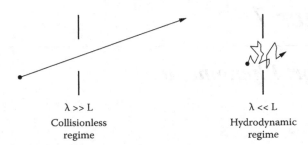

$\lambda \gg L$ $\lambda \ll L$
Collisionless   Hydrodynamic
regime   regime

Figure 7.1 Different regimes corresponding to different magnitudes of λ/L, the ratio of mean-free-path to an external length, in this case the size of the hole in the wall.

where

$$\lambda = (2mk_B T)^{-1}$$

$$C = n\,(2\pi m k_B T)^{-3/2} \tag{7.3}$$

We easily obtain the result

$$I = n\sqrt{\frac{k_B T}{2\pi m}} \tag{7.4}$$

Since the flux is proportional to v_x, fast atoms have a higher flux than slow ones, and so the escaped gas has a higher average energy per particle than the gas left behind. If the original volume contains a mixture of two gases at the same temperature, with atomic masses m_1, m_2, and densities n_1, n_2, then they effuse out of a small hole at different rates, with a ratio

$$\frac{I_1}{I_2} = \frac{n_1}{n_2}\sqrt{\frac{m_2}{m_1}} \tag{7.5}$$

This formula is the basis for a method to separate nuclear isotopes.

In the hydrodynamic regime, where $\lambda \ll L$, the gas reaches local thermal equilibrium over a distance small compared to L, but large compared to λ. It has well-defined local properties that vary slowly in space and in time:

$$T(\mathbf{r}, t) : \text{Local temperature}$$
$$n(\mathbf{r}, t) : \text{Local density} \tag{7.6}$$
$$\mathbf{u}(\mathbf{r}, t) : \text{Local flow velocity}$$

The flow trajectory with $\mathbf{u}$ as tangent vector is called a "streamline." In general the system settles into local thermodynamic equilibrium rather quickly, in the order of a collision time, but it takes much longer for the system to approach a uniform state. The local equilibrium is described by a local Maxwell–Boltzmann distribution

$$f(\mathbf{p}, \mathbf{r}, t) = Cn(\mathbf{r}, t)\exp\left[-\frac{|\mathbf{p}-m\mathbf{u}(\mathbf{r}, t)|^2}{2mk_B T(\mathbf{r}, t)}\right] \tag{7.7}$$

where the local variables are governed by the equations of hydrodynamics.

7.2   Maxwell's Demon

As we have noted, effusion is a velocity filter, but it acts in the same manner in both directions. A velocity filter that acts only in one direction would violate the second law of thermodynamics.

Maxwell imagined a way to do this by postulating a "demon" who operates a trap door in a wall separating two gases A and B, which were initially in equilibrium at the same temperature. The demon opens the door to allow fast atoms to go from A to B, but not the slow ones, and allow slow atoms to go from B to A, but not the fast ones. As time goes on, the average energy in A will rise, while that in B will fall. Thus the temperature of A will rise and that of B will fall "spontaneously." This fanciful idea has provoked much debate, centering on whether the demon should be considered part of the system, whether he/she/it has entropy, etc. Szilard (1929) pointed out that the demon needs information concerning the velocity of the approaching atoms, and that the second law can be preserved by regarding information as negative entropy. This idea has blossomed into the field of information theory, with applications to the theory of computation (Leff and Rex 1990).

7.3   Nonviscous Hydrodynamics

The hydrodynamic regime is based on the smallness of λ/L, where L refers to the characteristic wavelength of spatial variations. To the lowest order, we consider the limit $\lambda/L \to 0$, in which collisions are neglected, and the change in the local variables are governed solely by conservation laws. This leads to nonviscous hydrodynamics. In this limit the local equilibrium persists indefinitely; there is no damping mechanism for it to decay to a uniform state.

The relevant conservation laws are those for number of particle, and for momentum and energy. Because of particle conservation, the mass density

$$\rho(\mathbf{r}, t) = mn(\mathbf{r}, t) \tag{7.8}$$

satisfies the equation of continuity

$$\frac{\partial \rho}{\partial t} + \nabla \cdot (\rho \mathbf{u}) = 0 \quad \text{(continuity equation)} \tag{7.9}$$

The conservation of momentum is expressed through Newton's equation $F = ma$ in a local frame comoving with the gas along a streamline. Consider an element of the gas contained between x and $x + dx$, in a small cylinder of normal cross section A. In the absence of collisions, the x-component of the force acting on it due to neighboring gas elements arises purely from hydrostatic pressure:

$$dF_x = [P(x) - P(x + dx)]A = -A\frac{\partial P}{\partial x}dx \tag{7.10}$$

where $P(x)$ is the local pressure. This is the total net force on the element if we neglect collisions, which can create a shear force that leads to viscosity.

Newton's equation now states

$$-A\frac{\partial P}{\partial x}dx = \frac{du_x}{dt}dm \tag{7.11}$$

where the mass element dm is given by

$$dm = A\rho dx \tag{7.12}$$

Thus, we have

$$\rho\frac{du_x}{dt} + \frac{\partial P}{\partial x} = 0 \tag{7.13}$$

where du_x/dt is evaluated in the comoving frame. In a fixed frame in the laboratory, it is given by

$$\frac{du_x}{dt} = \frac{\partial u_x}{\partial t} + \left(u_x\frac{\partial}{\partial x} + u_y\frac{\partial}{\partial y} + u_z\frac{\partial}{\partial z}\right)u_x \tag{7.14}$$

Generalizing the above considerations to any component of $\mathbf{u}$, and adding an external force per unit volume $\mathbf{f}^{\text{ext}}$, we obtain *Euler's equation*

$$\rho\left(\frac{\partial}{\partial t} + \mathbf{u}\cdot\nabla\right)\mathbf{u} + \nabla P = \mathbf{f}^{\text{ext}} \quad \text{(Euler's equation)} \tag{7.15}$$

When collisions are ignored, there is no mechanism for energy transfer between a gas element and its neighbors. This means a gas element can only undergo adiabatic transformations in a comoving frame along a streamline. For an ideal gas we have

$$\left(\frac{\partial}{\partial t} + \mathbf{u}\cdot\nabla\right)(P\rho^{-\gamma}) = 0 \quad \text{(adiabatic condition)} \tag{7.16}$$

where $\gamma = C_P/C_V$, and the local equation of state gives $P = \rho k_B T/m$.

To be consistent with the premise $\lambda/L \to 0$, we must assume small deviations from equilibrium. Accordingly, we keep the local velocity $\mathbf{u}$, and all spatial and time derivatives, only to first order. Second order quantities, defined as products of first-order quantities, will be neglected. In particular we put

$$\nabla\cdot(\rho\mathbf{u}) = \rho\nabla\cdot\mathbf{u} + \mathbf{u}\cdot\nabla\rho$$

$$\approx \rho\nabla\cdot\mathbf{u} \tag{7.17}$$

because $\mathbf{u}\cdot\nabla\rho$ is of second-order smallness. This approximation leads to the linearized equations of nonviscous hydrodynamics:

$$\frac{\partial\rho}{\partial t} + \rho\nabla\cdot\mathbf{u} = 0$$

$$\rho\frac{\partial\mathbf{u}}{\partial t} + \nabla P = \mathbf{f}^{\text{ext}}$$

$$\frac{\partial}{\partial t}(P\rho^{-\gamma}) = 0 \tag{7.18}$$

where $P = \rho k_B T/m$.

7.4   Sound Wave

Differentiating the first hydrodynamic equation with respect to time, and keeping only
first-order terms, we obtain

$$\frac{\partial^2 \rho}{\partial t^2} + \rho \nabla \cdot \frac{\partial \mathbf{u}}{\partial t} = 0 \tag{7.19}$$

where a term $(\partial \rho / \partial t) \nabla \cdot \mathbf{u}$ has been neglected, because it is the product of two derivatives, and thus of second order. Substituting $\partial \mathbf{u} / \partial t$ from the second hydrodynamic
equation with $\mathbf{f}_{\text{ext}} = 0$, we have

$$\frac{\partial^2 \rho}{\partial t^2} - \rho \nabla \cdot \left(\frac{1}{\rho} \nabla \right) P = 0 \tag{7.20}$$

To first order, this is equivalent to

$$\frac{\partial^2 \rho}{\partial t^2} - \nabla^2 P = 0 \tag{7.21}$$

To evaluate $\nabla^2 P$, we make use of the third hydrodynamics equation:

$$\nabla^2 P = \nabla \cdot \nabla P = \nabla \cdot \left[\left(\frac{\partial P}{\partial \rho} \right)_S \nabla \rho \right]$$

$$\approx \left(\frac{\partial P}{\partial \rho} \right)_S \nabla^2 \rho = \rho \kappa_S \nabla^2 \rho \tag{7.22}$$

where κ_S is the adiabatic compressibility. Thus we have

$$\nabla^2 \rho - \frac{1}{c^2} \frac{\partial^2 \rho}{\partial t^2} = 0 \tag{7.23}$$

where

$$c = \frac{1}{\sqrt{\rho \kappa_S}} \tag{7.24}$$

This gives a wave equation for a sound wave of velocity c.

   Sound represents a collective motion of the atoms, and exists only when there are
collisions. The free sound wave here is obtained here by treating λ / L to zeroth order,
in which the effect of collisions produces a local thermal equilibrium. The next order
will lead to viscosity and damping.

7.5   Diffusion

We begin the consideration of first-order effects in λ / L by looking at how a density
gradient in the gas relaxes. The mechanism involved is the transport of particles from
one place to another.

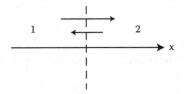

Figure 7.2   Diffusion results from a higher flux of particles in one direction than the other, due to a density gradient.

Suppose the density of a gas at points 1 and 2 along the x axis are n_1 and n_2 respectively. Atomic collisions tend to iron out the density variation, because there is a higher flux of particles going from the high density to the low density region, as indicated in Figure 7.2. This is the physical mechanism that gives rise to *diffusion,* which we discussed in Section 5.2 via the random walk.

We assume that the temperature is uniform, so that the atoms have a most probable velocity $\bar{v} = \sqrt{2k_B T/mT}$. In 3D space, on average one-sixth of the atoms travel along the positive x axis, and one sixth along the negative x axis. The flux of particles from 1 to 2 is therefore $n_1 \bar{v}/6$, that from 2 to 1 is $n_2\bar{v}/6$, and the net flux is from 1 to 2 is $(n_1 - n_2)\bar{v}/6$.

The densities at 1 and 2 can affect each other only if the separation is larger than a mean free path. Assume that they are separate by a distance r_0 of order of the mean free path. The x-component of the particle current density can be written as

$$j_x \approx -\frac{r_0 \bar{v}}{6} \frac{\partial n}{\partial x} \tag{7.25}$$

The minus sign occurs because a positive gradient along the x axis drives particles in the negative direction. In vector notation, we have

$$\mathbf{j} = -D\nabla n$$

$$D = \frac{r_0 \bar{v}}{6} \tag{7.26}$$

where D is the *diffusion constant.*

Since no atoms can be destroyed nor created, the particle current must satisfy the continuity equation

$$\frac{\partial n}{\partial t} + \nabla \cdot \mathbf{j} = 0 \tag{7.27}$$

Using $\mathbf{j}$ obtained earlier, we obtain the same diffusion equation obtained in Chapter 5 through the random walk:

$$\frac{\partial n(\mathbf{r}, t)}{\partial t} - D\nabla^2 n(\mathbf{r}, t) = 0 \tag{7.28}$$

With the initial condition that all N particles are located at $\mathbf{r} = 0$, the solution is

$$n(\mathbf{r}, t) = \frac{N}{(4\pi D)^{3/2}} \frac{e^{-r^2/(4\pi Dt)}}{t^{3/2}} \tag{7.29}$$

where $r = |\mathbf{r}|$. The initial condition is recovered in the limit

$$n(\mathbf{r}, t) \xrightarrow[t \to 0]{} N\delta^3(\mathbf{r}) \tag{7.30}$$

Conservation of particles is expressed through the fact

$$\int d^3r\, n(\mathbf{r}, t) = N \tag{7.31}$$

As time goes on, the particles diffuse out from the origin, forming a Gaussian distribution with an expanding width $\sqrt{4\pi Dt}$.

Using $\bar{v} = \sqrt{2k_B T/m}$, $r_0 = 2\lambda = 2/(n\sigma)$, where σ is the collision cross section, we obtain

$$D \approx \frac{1}{3n\sigma}\sqrt{\frac{2k_B T}{m}} \tag{7.32}$$

which gives an order-of-magnitude estimate.

Still other approaches to diffusion will be discussed via Browian motion (Section 10.5), and general stochastic processes (Section 11.5, Problem 12.6).

7.6   Heat Conduction

Assume now that the density n is uniform, while the temperature T varies slowly in space. The fluxes of particles in Figure 7.2 are now equal to $n\bar{v}/6$ in both directions. The average kinetic energy per particle is $lk_B T/2$, where l is the number of degrees of freedom. The heat flux from 1 to 2 is given by the flux of thermal energy

$$q_x = \frac{n\bar{v}l}{12}(k_B T_1 - k_B T_2) \approx -\frac{n\bar{v}r_0 lk}{12}\frac{\partial T}{\partial x} \tag{7.33}$$

We can write, in vector notation,

$$\mathbf{q} = -\kappa \nabla T$$

$$\kappa = \frac{1}{6}n\bar{v}r_0 c_V \tag{7.34}$$

where $c_V = lk/2$, and κ is a transport coefficient called the *coefficient of thermal conductivity*. Taking again $\bar{v} = \sqrt{2k_B T/m}$, $r_0 = 2\lambda = 2/(n\sigma)$, we have the order-of-magnitude estimate

$$\kappa = \frac{c_V}{3n\sigma}\sqrt{\frac{2k_B T}{m}} \tag{7.35}$$

When no work is performed, the heat absorbed by an element is equal to the increase in its internal energy, according to the first law of thermodynamics. This leads to the conservation law

$$\nabla \cdot \mathbf{q} + \frac{\partial u}{\partial t} = 0 \tag{7.36}$$

where u is the internal energy per unit volume. Using (7.34), we have

$$\frac{\partial u}{\partial t} - \kappa \nabla^2 T = 0 \tag{7.37}$$

For an ideal gas $u = nc_V T$, where c_V is the specific per particle. Thus we have a diffusion equation called the *heat conduction equation:*

$$\frac{\partial T}{\partial t} - \frac{\kappa}{nc_V} \nabla^2 T = 0 \tag{7.38}$$

7.7 Viscosity

The velocity of a gas flowing past a wall has a profile illustrated in Figure 7.3. Here, the gas is flowing along the x-direction with a nonuniform flow velocity $u_x(y)$. The gas sticks to the wall at $y = 0$, as expressed by the fact $u_x(0) = 0$. We assume that the density and temperature are uniform.

Consider a plane normal to the y axis, shown as the dotted line in Figure 7.3. The gas above this plane experiences a frictional force per unit area F_y given empirically by

$$F_y = -v \frac{\partial u_x(y)}{\partial y} \tag{7.39}$$

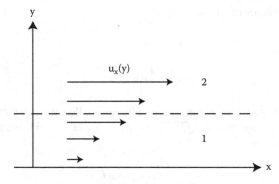

Figure 7.3 Particles sticking to the wall cause viscosity.

which defines v, the *coefficient of viscosity*. This force is caused by a transport of "x-component momentum" along the y-direction, from 1 to 2 in Figure 7.3. We now calculate v, using the definition

$$F_y = \text{Net flux of "x component momentum" along the } y \text{ direction} \qquad (7.40)$$

Note that u_x is the average collective flow velocity, and the transported momentum is that of the collective flow. The individual molecules, of course, dart about in all directions with average speed $\bar{v}$ relative to the flow velocity. Since the "x component momentum" per particle is mu_x, and the flux of particles in the y direction is $n\bar{v}/6$, we have

$$F_y = \frac{1}{6} n\bar{v}m[u_x(y_1) - u_x(y_2)] \qquad (7.41)$$

Choosing the points 1 and 2 to be separated by a distance r_0 of the order of a mean free path, we obtain

$$v = \frac{1}{6} n\bar{v}mr_0 \qquad (7.42)$$

Putting $r_0 = 2\lambda$, we have the estimate

$$v = \frac{\sqrt{2mk_B T}}{3\sigma} \qquad (7.43)$$

Note that this is independent of the density, a surprising prediction borne out by experiments on gases.

7.8 Navier–Stokes Equation

When viscosity is taken into account, the forces acting on a fluid element are no longer normal to the surfaces of the element. The hydrostatic pressure P is now generalized to a pressure tensor P_{ij}, which gives the jth component of the force per unit area

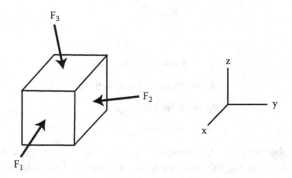

Figure 7.4 Due to viscosity, a force acting on a liquid element is not normal to the surface of the element, but has a shear component.

acting on the ith face, as illustrated in Figure 7.4. The vector force $\mathbf{F}_i$ on the ith face has the components indicated below:

$$\mathbf{F}_1 = A(P_{11}, P_{12}, P_{13})$$

$$\mathbf{F}_2 = A(P_{21}, P_{22}, P_{23})$$

$$\mathbf{F}_{3i} = A(P_{31}, P_{32}, P_{33}) \tag{7.44}$$

where A is the surface area. We can thus split the pressure tensor into two terms:

$$P_{ij} = \delta_{ij} P + P'_{ij} \tag{7.45}$$

where the off-diagonal term P'_{ij} depends on the viscosity, and turns out to have the form (Huang 1987)

$$P'_{ij} = -\nu \left[\frac{\partial u_i}{\partial x_j} + \frac{\partial u_j}{\partial x_i} - \frac{2}{3}\delta_{ij}\nabla \cdot \mathbf{u} \right] \tag{7.46}$$

The generalization of Euler's equation reads

$$\rho \left(\frac{\partial}{\partial t} + \mathbf{u} \cdot \nabla \right) u_i + \frac{\partial P_{ij}}{\partial x_j} = f_i^{\text{ext}} \tag{7.47}$$

or, more explicitly,

$$\rho \left(\frac{\partial}{\partial t} + \mathbf{u} \cdot \nabla \right) \mathbf{u} + \nabla \left(P - \frac{\nu}{3}\nabla \cdot \mathbf{u} \right) - \mu\nabla^2\mathbf{u} = \mathbf{f}^{\text{ext}} \tag{7.48}$$

where $\mathbf{f}^{\text{ext}}$ is the external force per unit volume. This is known as the *Navier–Stokes equation*.

The existence of viscosity furnishes a new scale in hydrodynamics. The dimensionality of viscosity is

$$\nu \sim \frac{\text{force/area}}{\text{velocity/length}} \sim \frac{mu}{tL^2}\frac{L}{u} \sim \frac{m}{Lt} \tag{7.49}$$

where m = mass, u = velocity, L = length, t = time. Using ν we can define a dimensionless quantity characterizing hydrodynamic flow called the *Reynolds number*:

$$R = \frac{\rho L u_0}{\nu} \tag{7.50}$$

where

$$\rho = \text{mass density}$$

$$L = \text{characteristic length}$$

$$u_0 = \text{flow velocity}$$

$$\nu = \text{viscosity} \tag{7.51}$$

for a stationary object of size L, immersed in a fluid of mass density ρ, and viscosity ν, flowing with velocity u_0, this number marks the onset of turbulence. When $R \ll 1$ the fluid flows past the object in streamline flow, and when $R \gg 1$ we have turbulent flow.

Problems

7.1 A high-vacuum chamber develops a small crack of area σ, and air from the outside leaks in by effusion.

(a) Find the rate of air molecules leaking in through the crack.

(b) After a short time, the leak was discovered, and patched up by diligent students. The small amount of gas came to equilibrium inside the chamber. Show that its absolute temperature is higher than that of the air outside by a factor 4/3.

Hint: Suppose the leak existed for a time τ. Calculate the energy E and number of molecules N that got through during that time. The equilibrium temperature inside the chamber is determined by the ratio E/N.

7.2 Natural uranium ore contains isotopes ^{238}U and ^{235}U with abundances 99.27% and 0.73%, respectively. To increase the relative abundance of ^{235}U, a sample of natural uranium is vaporized, and made to effuse successively into a series of vacuum chambers. How many stages are required to achieve equal abundance in the two isotopes?

7.3 Show that the velocity of sound of an ideal gas is

$$c = \sqrt{\frac{k_B T \gamma}{m}}$$

where $\gamma = C_P/C_V$. Evaluate this for air (nitrogen) at STP.

7.4 Sound propagates adiabatically because the effect of heat conduction can be neglected. Verify this in a real gas, as follows.

Consider a sound wave of wavelength L and period $\tau = L/c$, where c is the sound velocity. Let ΔT be the variation in temperature over L. Then the magnitude of the heat flux is $q \approx K \Delta T/L$, where K is the coefficient of thermal conductivity. The amount of heat transferred by conduction across unit area over the distance L is therefore $Q_1 = q\tau = K \Delta T/c$. The amount of heat needed to equalize the temperature is $Q_2 = C_P \Delta T$. Thus, heat conductivity may be ignored if $Q_1 \ll Q_2$.

Test this condition for a sound wave with $L = 10$ ft in air at STP, with the following data:

$$\rho = 0.08 \text{ lb ft}^{-3}$$

$$c = 1088 \text{ ft s}^{-1}$$

$$C_P = 0.24 \text{ Btu lb}^{-1} \, (^\circ\text{F})^{-1}$$

$$K = 0.0157 \text{ Btu hr}^{-1}\text{ft}^{-1}(^\circ\text{F})^{-1}$$

7.5 Rederive the equation for a sound wave, using the Navier–Stokes equation instead of the Euler equation. Find the damping coefficient for sound.

7.6 A long thin tube along the x axis contains a gas of N particles initially concentrated at $x = 0$. A particle detector is placed at $x = L$. At what time does the detector register the first signal?

7.7 A gas is sealed between the thermal panes of a window. The insulating power of the window is taken to be the inverse of the coefficient of thermal conductivity of the gas. Normally the gas is air, of average molecular weight 30. What should be the molecular weight of the gas, in order to double the insulating power?

Hint: Consider how the coefficient of thermal conductivity depends on molecular weight, through the mass of the molecule and the collision cross section.

7.8 Heat is generated uniformly in the Earth's interior due to radioactivity, at the rate W cal/g. Assume that the Earth is spherically symmetric with radius R, uniform mass density ρ, and coefficient of thermal conductivity κ. Ignore the effect of all other heat sources.

Find the temperature $T(r)$, as a function of distance r from the center of the Earth.

7.9 During heat transfer characterized by the heat flux vector $\mathbf{q}$, there is both entropy flow and irreversible entropy production.

(a) Show that the entropy density s satisfies the equation

$$\frac{\partial s}{\partial t} + \frac{1}{T} \nabla \cdot \mathbf{q} = 0$$

where T is the absolute temperature.

(b) Show that the rate of entropy production is given by $R_s = \mathbf{q} \cdot \nabla (1/T)$, hence

$$R_s = \kappa \left(\frac{\nabla T}{T} \right)^2$$

7.10 The temperature of a pond is just above freezing. The temperature of the air suddenly drops by ΔT, and a sheet of ice begins to form at the surface of the pond, and thickens as time goes on. Find the rate at which the thickness of the ice sheet grows.

Consider only the transition of water to ice, and ignore the cooling of the ice once it is formed. Let the latent of ice be ℓ, the mass density of water be ρ, and the coefficient of thermal conductivity of ice be κ.

References

Huang, K., *Statistical Mechanics*, 2nd ed., Wiley, New York, 1987, Section 5.8.

Leff, H.S. and A.F. Rex (eds.), *Maxwell's Demon: Entropy, Information, Computing*, Princeton University Press, Princeton, 1990.

Szilard, L., *Z. F. Physik*, **53**:840 (1929).

Chapter 8

Canonical Ensemble

8.1 Review of the Microcanonical Ensemble

The time has come to look beyond the ideal gas, and consider systems with interactions: a dense gas, a liquid, a solid. We already have the principle for doing this, for the microcanonical ensemble defined in Section 5.7 can be applied to any system, not just the ideal gas.

All members of the microcanonical ensemble have the same energy E, within a small tolerance. The entropy of the system is given by

$$S(E) = k_B \ln \Gamma(E) \tag{8.1}$$

where k_B is Boltzmann's constant, and $\Gamma(E)$ is the total number of states of the system at energy E. The dependence on the number of particles and total volume have been left understood. The internal energy is simply E, and the absolute temperature T defined by

$$\frac{1}{T} = \frac{\partial S(E)}{\partial E} \tag{8.2}$$

This method is very convenient when we deal with discrete degrees of freedom, for the calculation of $\Gamma(E)$ is a matter of combinatorics, as illustrated by problems in this chapter. For systems like a real gas, which is specified through a Hamiltonian, the microcanonical ensemble is not as easy to use, and we seek a more convenient route.

8.2 Classical Canonical Ensemble

The key to a new method is to relax the requirement that the energy be fixed, by allowing the system to exchange energy with the environment, which we call the heat reservoir. Now the reservoir and the system form an isolated system, which can be treated in the microcanonical ensemble. Thus, what is new is a change in emphasis, and no new principles are needed.

Consider an isolated system divided into two subsystems: a "large" one regarded as a heat reservoir, and a "small" one on which we focus our attention. Let us label

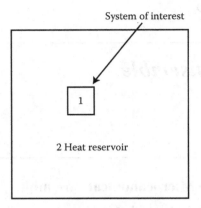

System of interest

1

2 Heat reservoir

Figure 8.1 We focus attention on a "small" system 1, which is part of a larger system. The rest of the sytem 2 acts as a heat reservoir for system 1.

the small system 1, and the heat reservoir 2, as illustrated schematically in Figure 8.1. In the end, of course, even the small system will be made large, to approach the thermodynamic limit. The canonical ensemble is the ensemble of the small system.

Neglecting the interactions across the boundary between the two systems, we take the total Hamiltonian to be the sum

$$H(p_1, q_1, p_2, q_2) = H_1(p_1, q_1) + H_2(p_2, q_2) \tag{8.3}$$

where $\{p_1, q_1\}, \{p_2, q_2\}$ respectively denote the momenta and coordinates of the two subsystems. The total number of particles N, and total energy E, are given by

$$N = N_1 + N_2$$

$$E = E_1 + E_2 \tag{8.4}$$

We keep N_1 and N_2 separately fixed, but allow E_1 and E_2 to fluctuate. In other words, the dividing walls between the subsystem allow energy exchange, but not particle exchange. We assume that system 1 is infinitesimally small in comparison with system 2, even though both are macroscopic systems:

$$N_2 \gg N_1$$

$$E_2 \gg E_2 \tag{8.5}$$

The phase-space volume occupied by system 2 is given by

$$\Gamma_2(E_2) = \int_{E_2} dp_2 dq_2 \tag{8.6}$$

where the subscript E_2 is shorthand for the condition

$$E_2 < H_2(p_2, q_2) < E_2 + \Delta \tag{8.7}$$

We are interested in the probability of occurrence of the microstate $\{p_1, q_1\}$ of subsystem 1, regardless of the state of system 2:

$$\text{Probability that 1 is in } dp_1 dq_1 \propto dp_1 dq_1 \Gamma_2(E - E_1) \qquad (8.8)$$

This gives the distribution function $\rho_1(p_1, q_1)$ of system 1 in its own Γ space:

$$\rho_1(p_1, q_1) = \Gamma_2(E - E_1)$$
$$= \Gamma_2(E - H_1(p_1, q_1)) \qquad (8.9)$$

Since $E_1 \ll E$, we expand the above in powers of E_1 to lowest order. It is convenient to expand the logarithm of Γ_2, which can be expressed in terms of the entropy of system 2:

$$k \ln \Gamma_2(E - E_1) = S_2(E - E_1)$$
$$= S_2(E) - E_1 \left. \frac{\partial S_2(E')}{\partial E'} \right|_{E'=E} + \cdots$$
$$\approx S_2(E) - \frac{E_1}{T} \qquad (8.10)$$

where T is the temperature of system 2. This relation becomes exact in the limit when system 2 becomes a heat reservoir. The temperature of the heat reservoir T fixes the temperature of system 1.

The density function for system 1 is therefore

$$\rho_1(p_1, q_1) = e^{S_2(E)/k} e^{-E_1/k_B T} \qquad (8.11)$$

The first factor is a constant, which can be dropped by redefining the normalization. In the second factor the energy of the system can be replaced by its Hamiltonian. Thus

$$\rho_1(p_1, q_1) = e^{-\beta H_1(p_1, q_1)} \qquad (8.12)$$

with

$$\beta = \frac{1}{k_B T} \qquad (8.13)$$

This is the *canonical ensemble*, appropriate for a system of a fixed number of particles, in contact with a heat reservoir of temperature T. Since we shall refer only to the small system from now on, the subscript 1 is unnecessary, and will be omitted. Schematic representations of the microcanonical and canonical ensembles are given in Figure 8.2.

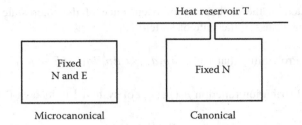

Figure 8.2 Schematic representations of the microcanonical ensemble and canonical ensemble.

8.3 The Partition Function

The *partition function* for the canonical ensemble is defined by

$$Q_N(T, V) \equiv \int \frac{dpdq}{\tau} e^{-\beta H(p,q)}$$

$$\tau = N!h^{3N} \qquad (8.14)$$

where $dp = d^{3N}p$, $dq = d^{3N}q$, and each d^3q integration ranges over the volume V of the system. The factor τ makes Q_N dimensionless by supplying the unit h for an elementary cell in phase space. The factor $N!$ comes from correct Boltzmann counting, which takes into account the indistinguishability of particles in quantum mechanics. It is included here because we assume we are dealing with atoms. The factor $N!$ should be omitted if we have a collection of distinguishable objects.

As we shall see, all thermodynamic information can be obtained from the partition function. In particular, the entropy is completely determined, with no arbitrary constants, and the factor $N!$ will make it an extensive quantity.

8.4 Connection with Thermodynamics

The thermodynamic internal energy is the ensemble average of the energy:

$$U = \langle H \rangle = \frac{\int dpdq H e^{-\beta H(p,q)}}{\int dpdq e^{-\beta H(p,q)}} \qquad (8.15)$$

This can be obtained from the partition function by differentiation with respect to β:

$$U = -\frac{\partial}{\partial \beta} \ln Q_N \qquad (8.16)$$

The general connection between the canonical ensemble and thermodynamics is given by the statement

$$Q_N(T, V) = e^{-\beta A(V,T)} \tag{8.17}$$

where is $A(V, T)$ is the Helmholtz energy.

To show that Equation (8.17) correctly identifies the free energy, let us rewrite it in the form

$$\int \frac{dpdq}{\tau} e^{-\beta H(p,q)} = e^{-\beta A(V,T)}$$

$$\int \frac{dpdq}{\tau} e^{\beta[A(V,T)-H(p,q)]} = 1 \tag{8.18}$$

Differentiating both sides with respect to β, we obtain

$$\int \frac{dpdq}{\tau} e^{\beta[A(V,T)-H(p,q)]} \left[A(V, T) + \beta \frac{\partial A(V, T)}{\partial \beta} - H(p, q) \right] = 0 \tag{8.19}$$

The first two terms are not functions of p, q and can be taken outside of the integral, which is equal to 1. The second term gives the internal energy. Thus

$$A(V, T) + \beta \frac{\partial A(V, T)}{\partial \beta} - U = 0$$

$$A(V, T) - T \frac{\partial A(V, T)}{\partial T} - U = 0 \tag{8.20}$$

This is consistent with the thermodynamic relations

$$S = -\frac{\partial A(V, T)}{\partial T}$$

$$A = U - TS \tag{8.21}$$

Therefore we have a correct definition of $A(V, T)$, from which we can obtain all thermodynamic functions through the Maxwell relations.

8.5 Energy Fluctuations

The systems in the canonical ensemble have different energies, whose values fluctuate about the mean energy U determined by the temperature. To calculate the mean-square fluctuation of the energy we start with the expression

$$U = -\frac{\partial}{\partial \beta} \ln \int dpdq e^{-\beta H} \tag{8.22}$$

and differentiate with respect to β:

$$\frac{\partial U}{\partial \beta} = -\frac{\int dpdq H^2 e^{-\beta H}}{\int dpdq e^{-\beta H}} + \frac{\left(\int dpdq H e^{-\beta H}\right)^2}{\left(\int dpdq e^{-\beta H}\right)^2}$$

$$= -\langle H^2 \rangle + \langle H \rangle^2 \qquad (8.23)$$

Using thermodynamic definitions, we can rewrite

$$\frac{\partial U}{\partial \beta} = \frac{\partial U}{\partial T}\frac{\partial T}{\partial \beta} = -k_B T^2 \frac{\partial U}{\partial T} = -k_B T^2 C_V \qquad (8.24)$$

Thus

$$\langle H^2 \rangle - \langle H \rangle^2 = k_B T^2 C_V \qquad (8.25)$$

For macroscopic systems the left side is of order N^2, while the right side is of order N. Energy fluctuations are therefore "normal," and become negligible when $N \to \infty$. This is why the results of the canonical and microcanonical ensembles coincide in that limit.

8.6   Minimization of Free Energy

We have learned in thermodynamics that a system at fixed V, T will seek the state of minimum free energy. This principle can be derived using the canonical ensemble.

Note that the partition function [Equation (8.14)] is an integral over states, but the integrand depends only on the Hamiltonian, whose value is the energy. We can convert the integral into one over the energy by writing

$$Q_N(V, T) = \int dE\, e^{-\beta E} \int \frac{dpdq}{\tau} \delta\left(H(p, q) - E\right) \qquad (8.26)$$

This is a trivial rewriting, for the dE integration can be done by simply setting $E = H(p, q)$. The $dpdq$ integral now gives the density of states at energy E:

$$\omega(E) = \int \frac{dpdq}{\tau} \delta\left(H(p, q) - E\right) \qquad (8.27)$$

which is related to the entropy by

$$S(E) = k_B \ln \omega(E) \qquad (8.28)$$

Thus

$$Q_N(V, T) = \int dE\, \omega(E)\, e^{-\beta E} = \int dE\, e^{-\beta[E - TS(E)]} \qquad (8.29)$$

or

$$Q_N(V, T) = \int dE \, e^{-\beta A(E)} \tag{8.30}$$

where $A(E) \equiv E - TS(E)$.

The dominant contribution to the integral will come from the value of E for which $A(E)$ is minimum. This can be seen as follows. Let the minimum of $A(E)$ occur at $\bar{E}$, which is determined by the condition

$$\left. \frac{\partial A}{\partial E} \right|_{E=\bar{E}} = \left[1 - T \frac{\partial S}{\partial E} \right]_{E=\bar{E}} = 0 \tag{8.31}$$

This gives

$$\left. \frac{\partial S}{\partial E} \right|_{E=\bar{E}} = \frac{1}{T} \tag{8.32}$$

which is only the thermodynamic relation between entropy and temperature. The second derivative of $A(E)$ gives

$$\frac{\partial^2 A}{\partial E^2} = -T \frac{\partial^2 S}{\partial E^2} \tag{8.33}$$

From the Maxwell relation $\partial E / \partial S = -T$, we have $\partial S / \partial E = -1/T$. Hence

$$\frac{\partial^2 S}{\partial E^2} = \frac{1}{T^2} \frac{\partial T}{\partial E} = \frac{1}{T^2 C_V} \tag{8.34}$$

Thus the second derivative is positive:

$$\frac{\partial^2 A}{\partial E^2} = \frac{1}{T C_V} \tag{8.35}$$

showing that the free energy at $\bar{E}$ has a minimum (and not maximum). Now expand $A(E)$ about the minimum:

$$A(E) = A(\bar{E}) + \frac{1}{2 T C_V} (E - \bar{E})^2 + \cdots \tag{8.36}$$

Neglecting the higher-order terms, we have

$$Q_N(V, T) = e^{-\beta A(\bar{E})} \int dE \, e^{-(E-\bar{E})^2 / (2k_B T^2 C_V)} \tag{8.37}$$

In the thermodynamic limit C_V becomes infinite, and the integrand is very sharply peaked at $E = \bar{E}$, as illustrated in Figure 8.3. This is why we can neglect the higher-order terms, and evaluate the integral by extending the limits of integration from $-\infty$ to ∞, to obtain

$$Q_N = \sqrt{2\pi k_B T^2 C_V} \, e^{-\beta A(\bar{E})}$$

$$\ln Q_N(V, T) = -\frac{A(\bar{E})}{k_B T} + \frac{1}{2} \ln \left(2\pi k_B T^2 C_V \right) \tag{8.38}$$

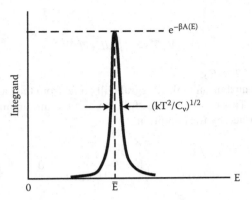

Figure 8.3 When the partition function is expressed as an integral over energy, the integrand is sharply peaked $E = \bar{E}$ corresponding to a minimum in the free enegy $A = E - TS$.

In the thermodynamic limit, the first term is of order N, while the second term is of order $\ln N$, and can be neglected. Thus, we see that the system seeks a state of minimum free energy.

8.7 Classical Ideal Gas

We illustrate the classical canonical ensemble with the ideal gas. The Hamiltonian is given by

$$H = \sum_{i=1}^{N} \frac{\mathbf{p}_i^2}{2m} \tag{8.39}$$

The partition function is

$$Q_N(T, V) = \int \frac{d^{3N}p\, d^{3N}q}{N! h^{3N}} e^{-\beta(\mathbf{p}_1^2 + \cdots + \mathbf{p}_N^2)/2m} \tag{8.40}$$

$$= \frac{V^N}{N!} \left(\int_{-\infty}^{\infty} \frac{dp}{h} e^{-\beta p^2/2m} \right)^{3N} = \frac{1}{N!} \left(\frac{V}{\lambda^3} \right)^N \tag{8.41}$$

where $\lambda = \sqrt{2\pi\hbar^2/mk_BT}$ is the thermal wavelength. Thus, the free energy is given by

$$A(V, T) = -k_BT \ln Q_N(T, V)$$

$$= Nk_BT \left[\ln\left(n\lambda^3\right) - 1 \right] \tag{8.42}$$

where $n = N/V$, and we have used the Stirling approximation $\ln N! \approx N \ln N - N$.

The chemical potential is given by the Maxwell relation

$$\mu = \left(\frac{\partial A}{\partial N}\right)_{V,T} = k_B T \ln\left(n\lambda^3\right) \tag{8.43}$$

and the entropy is given by $S = -\partial A/\partial T$, which leads to

$$S = -\left(\frac{\partial A}{\partial T}\right)_V = Nk\left[\frac{5}{2} - \ln(n\lambda^3)\right] \tag{8.44}$$

This is the *Sacher-Tetrode equation* that fixes the defects in the thermodynamic result [Equation (2.48)]. It also supplies the undetermined constant in Equation (6.31).

Problems

8.1 A perfect crystal has N lattice sites and M interstitial locations. An energy Δ is to remove an atom from a site and place it in an interstitial, when the number n of displaced atoms is much smaller than N or M.

(a) How many ways are there of removing n atoms from N sites?

(b) How many ways are there of placing n atoms on M interstitials?

(c) Use the microcanonical ensemble to calculate the entropy as a function of total energy E, and define the temperature.

*(d) Show that the average number of displaced atoms n at temperature T is given through

$$\frac{n^2}{(N-n)(M-n)} = e^{-\Delta/k_B T}$$

Obtain n for $\Delta \gg k_B T$, and $\Delta \ll k_B T$.

(e) Use this model for defects in a solid. Set $N = M$, and $\Delta = 1\,eV$. find the defect concentration at $T = 1000$ K and 300 K,

8.2 A one-dimensional chain, fixed at one end, is made of N identical elements each of length a. The angle between successive elements can be either $0°$ or $180°$, as shown in the accompanying sketch. There is no difference in the energies of these two possibilities. We can think of each element as either pointing right $(+)$ or left $(-)$. Suppose $N_\pm$ is the number of $\pm$ elements, and L is the total length of the chain. We have

$$N = N_+ + N_-$$

$$L = a\left(N_+ - N_-\right)$$

An interesting feature of this model is that the internal energy does not depend on L, but, as is to be shown below, there is a tension τ defined through

$$dU = TdS + \tau dL$$

It arises statistically, through the fact that a shorter chain can sample more of a phase space.

Find the following quantities, using the microcanonical ensemble:

(a) the entropy as a function of N and N_+

(b) the free energy as a function of N and N_+

(c) the tension τ as a function of T, N, L

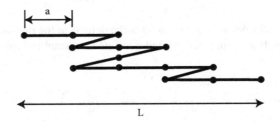

8.3 A chain made of N massless segments of equal length a hangs from a fixed point. A mass m is attached to the other end under gravity. Each segment can be in either of two states, up or down, as illustrated in the sketch. The segments have no mass, and the chain can go as far up as it can; there is no ceiling.

(a) Show that the partition function at temperature T is given by

$$Q_N = \left(1 + e^{-2mga/k_BT}\right)^N$$

(b) Find the entropy of the chain.

(c) Find the internal energy, and determine the length of the chain.

(d) Show that the chain obeys Hooke's law, namely, a small force pulling on the chain increases its length proportionately. Find the proportionality constant.

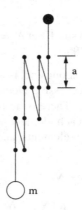

8.4 The unwinding of a double-stranded DNA molecule is like unzipping a zipper. The DNA has N links, each of which can be in one of two states: a closed state with energy 0, and an open state with energy Δ. A link can open only if all the links to its left are already open, as illustrated in the sketch.

(a) Show that the partition function of the DNA chain is

$$Q_N = \frac{1 - e^{-(N+1)\Delta/k_B T}}{1 - e^{-\Delta/k_B T}} \tag{8.45}$$

(b) Find the average number of open links in the low-temperature limit $k_B T \ll \Delta$.

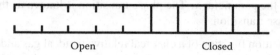

Open   Closed

8.5 Consider a piece of two-dimensional graphite where N carbon atoms form a honeycomb lattice. Assume that it costs energy Δ to remove a carbon atom from a lattice site and place it in the center of a hexagon to form a vacancy and an interstitial, as shown in the sketch.

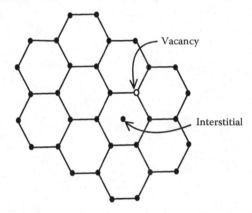

Vacancy

Interstitial

(a) Show that there are $N/2$ possible locations for interstitials.

(b) Consider a microcanonical ensemble of the system, at given total energy E. For M interstitials, find the statistical entropy for large N and M.

(c) Find the most probable value $\bar{M}$ of M, using the method of Lagrange multipliers.

(d) Find the equilibrium entropy S, and express the Lagrange multiplier in term of the temperature, defined by $T^{-1} (\partial S/\partial E)$.

(e) Given $\bar{M}$ in the limits $T \to 0$ and $T \to \infty$.

8.6 A particle can exist in only three states labeled by $n = 1, 2, 3$. The energies ϵ_n of these states depend on a parameter $x \geq 0$, with two of the energies degenerate:

$$\epsilon_1 = \epsilon_2 = bx^2 - \frac{1}{2}cx$$

$$\epsilon_3 = bx^2 + cx$$

where b and c are constants.

(a) Find the Helmholtz free energy per particle $a(x, T) = A_N(x, T)/N$ for a collection of N such particles, assuming that there are no inter-particle interactions.

(b) If x is allowed to freely vary at constant T, it will assume an equilibrium value $\bar{x}$ that minimizes the free energy. Find $\bar{x}$ as a function of T. Show that there is a phase transition, and find the transition temperature. Assume that $\bar{x}$ is small, and expand $\partial a(x, T)/\partial x$ in a power series in x to order x^2.

This model can be used to describe ions in a crystal subject to a uniform strain characterized by the parameter x. The phase transition is known as the "cooperative Jahn–Teller" phase transition.

8.7 Set up the partition function of a classical relativistic ideal gas and obtain the free energy in terms of an integral In the nonrelativistic and the ultra-relativistic limits, show that the chemical potentials are given by

Nonrelativistic: $\mu \approx mc^2 + k_B T \ln(n\lambda^3)$ $(\lambda = \sqrt{2\pi\hbar^2/mk_B T})$

Ultra-relativistic: $\mu \approx k_B T \ln(nL^3)$ $(L = \pi^{2/3}\hbar c/k_B T)$

Chapter 9

Grand Canonical Ensemble

9.1 The Particle Reservoir

The grand canonical ensemble is built upon the canonical ensemble by relaxing the restriction to a definite number of particles. The relative probability of finding the system with N particles at temperature T, in a volume V, is taken to be

$$\rho(N, V, T) = z^N Q_N(V, T) \tag{9.1}$$

where $z = e^{\beta \mu}$ is the fugacity. The chemical potential μ is a given external parameter, in addition to the temperature. The system exchanges energy with a heat reservoir of temperature T, which determines the average energy, and it exchanges particles with a "particle reservoir" of chemical potential μ, which determines the average number of particles. A schematic representation of the ensemble is shown in Figure 9.1.

The grand canonical ensemble is a more realistic representation of physical systems than the canonical ensemble, for we can rarely fix the total number of particles in a macroscopic system. A typical example is a volume of air in the atmosphere. The particle reservoir in this case is the rest of the atmosphere. The number of air molecules in the volume considered fluctuates about a mean value determined by the rest of the atmosphere.

9.2 Grand Partition Function

The ensemble average in the grand canonical ensemble is obtained by averaging the canonical average over the number of particles N. For example, the internal energy is given by

$$U = \frac{\sum E_N z^N Q_N}{\sum z^N Q_N} \tag{9.2}$$

where E_N is the average energy in the canonical ensemble:

$$E_N = -\frac{\partial}{\partial \beta} \ln Q_N \tag{9.3}$$

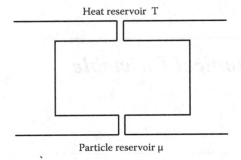

Heat reservoir T

Particle reservoir μ

Figure 9.1 Schematic representation of the grand canonical ensemble. The system exchanges particles within a particle reservoir with fixed chemical potential, and exchanges energy with a heat reservoir with fixed temperature.

It is useful to introduce the *grand partition function*:

$$Q(z, V, T) = \sum_{N=0}^{\infty} z^N Q_N(V, t) \tag{9.4}$$

in terms of which we can write

$$U = -\frac{\partial}{\partial \beta} \ln Q(z, V, T) \tag{9.5}$$

In the thermodynamic limit $V \to \infty$, we expect that

$$\frac{1}{V} \ln Q(z, V, T) \xrightarrow[V \to \infty]{} \text{Finite limit} \tag{9.6}$$

9.3 Number Fluctuations

The average number of particles is given by

$$\bar{N} = \frac{\sum N z^N Q_N}{\sum z^N Q_N} = z \frac{\partial}{\partial z} \ln Q(z, V, T) \tag{9.7}$$

Specifying $\bar{N}$ determines the chemical potential $\mu = k_B T \ln z$. The mean-square fluctuation can be obtained by differentiating again with respect to z:

$$z \frac{\partial}{\partial z} z \frac{\partial}{\partial z} \ln Q(z, V, T) = \frac{\sum N^2 z^N Q_N}{\sum z^N Q_N} - \left[\frac{\sum N z^N Q_N}{\sum z^N Q_N} \right]^2$$

$$= \overline{N^2} - \overline{N}^2$$

In terms of the chemical potential we can write

$$z\frac{\partial}{\partial z} = z\frac{\partial\mu}{\partial z}\frac{\partial}{\partial\mu} = k_B T\frac{\partial}{\partial\mu} \tag{9.8}$$

Thus

$$\overline{N^2} - \overline{N}^2 = (k_B T)^2\frac{\partial^2}{\partial\mu^2}\ln\mathcal{Q}(z, V, T) \tag{9.9}$$

Dividing both sides by V^2, we have the density fluctuation

$$\overline{n^2} - \overline{n}^2 = \frac{(k_B T)^2}{V^2}\frac{\partial^2}{\partial\mu^2}\ln\mathcal{Q}(z, V, T) \tag{9.10}$$

Assuming Equation (9.6), we see that this vanishes like V^{-1} in the thermodynamic limit, and makes the grand canonical ensemble equivalent to the canonical ensemble.

9.4   Connection with Thermodynamics

Assuming that the number fluctuation is vanishingly small, we need to keep only the largest term in the sum over N:

$$\ln\mathcal{Q}(z, V, T) = \ln\sum_{N=0}^{\infty} z^N Q_N(V, T) \approx \ln\left[z^{\bar{N}} Q_{\bar{N}}(V, T)\right]$$

$$= \bar{N}\ln z + \ln Q_{\bar{N}}(V, T) = \frac{\bar{N}\mu}{k_B T} + \ln Q_{\bar{N}}(V, T) \tag{9.11}$$

where $\bar{N}$ is the average number of particles. Now put

$$Q_N(V, T) = e^{-\beta A_N(V, T)}$$

$$A_N(V, T) = Na(v, T)$$

where $a(v, T)$ is the free energy per particle, and $v = V/N$. Then we have

$$\ln\mathcal{Q}(z, V, T) = \frac{\bar{N}}{k_B T}[\mu - a(\bar{v}, T)] \tag{9.12}$$

where $\bar{v} = V/\bar{N}$.

The pressure is given by a Maxwell relation:

$$P = -\left[\frac{\partial A_N(V, T)}{\partial V}\right]_{N,T} = -\frac{\partial}{\partial V}[Na(v, T)]_{N,T}$$

$$= -N\frac{\partial a(v, T)}{\partial V} = -\frac{\partial a(v, T)}{\partial v} \tag{9.13}$$

The chemical potential is given by

$$\mu = \left[\frac{\partial A_N(V, T)}{\partial N} \right]_{V,T} = a(v, T) + N \frac{\partial a(v, T)}{\partial N} \tag{9.14}$$

Since

$$\frac{\partial a(v, T)}{\partial N} = \frac{\partial a}{\partial v} \frac{\partial v}{\partial N} = -\frac{V}{N^2} \frac{\partial a}{\partial v} \tag{9.15}$$

we have

$$\mu = a(v, T) - v \frac{\partial a(v, T)}{\partial v} \tag{9.16}$$

Thus we obtain the thermodynamic relation

$$\mu = a(v, T) + Pv \tag{9.17}$$

Using Equation (9.12) we obtain

$$\ln \mathcal{Q}(z, V, T) = \frac{PV}{k_B T} \tag{9.18}$$

From now on, we shall omit the overhead bar in $\bar{N}$, and denote the average particle number by N, whenever there is no danger of confusion.

9.5   Parametric Equation of State and Virial Expansion

The pressure obtained from Equation (9.18) is a function of temperature and fugacity z. Combining it with Equation (9.7), we have the equation of state in parametric form:

$$\frac{P}{k_B T} = \frac{1}{V} \ln \mathcal{Q}(z, V, T)$$

$$n = \frac{1}{V} z \frac{\partial}{\partial z} \ln \mathcal{Q}(z, V, T) \tag{9.19}$$

The usual equation of state, where P is given as a function of T and density n, can be obtained by eliminating z.

In the gas phase of the system, that is, at sufficiently high temperatures and low densities, it is usually possible to expand the right sides as power series in z:

$$\frac{P}{k_B T} = \frac{1}{\lambda^3} \sum_{\ell=1}^{\infty} b_\ell z^\ell$$

$$\frac{N}{V} = \frac{1}{\lambda^3} \sum_{\ell=1}^{\infty} \ell b_\ell z^\ell \tag{9.20}$$

where b_ℓ are coefficients known as "cluster integrals," which generally depend on T, and $b_1 \equiv 1$ by definition. The thermal wavelength $\lambda = \sqrt{2\pi\hbar^2/mk_BT}$ is introduced purely as a convenient scale parameter. The ℓth cluster integral expresses the correlation among a cluster of ℓ particles due to interactions, and in the classical ideal gas they vanish for $\ell > 2$. By eliminating z, we can obtain P as a power series in n:

$$\frac{PV}{Nk_BT} = 1 + a_2\lambda^3 n + a_3(\lambda^3 n)^2 + \cdots \tag{9.21}$$

This is called the "virial expansion," and a_ℓ is called the ℓth "virial coefficient." Historically, the earliest information on the interparticle potential was obtained from experimental data on the second virial coefficient a_2.

9.6 Critical Fluctuations

We can now express the density fluctuation in terms of measurable thermodynamic coefficients. From Equations (9.10) and (9.18), we have

$$\overline{n^2} - \overline{n}^2 == \frac{k_BT}{V}\frac{\partial^2 P}{\partial\mu^2}$$

From Equation (9.17) we obtain

$$\frac{\partial\mu}{\partial v} = -v\frac{\partial^2 a(v)}{\partial v^2} = v\frac{\partial P}{\partial v} \tag{9.22}$$

where the dependence on T is left understood. Thus

$$\frac{\partial P}{\partial\mu} = \frac{\partial P}{\partial v}\frac{\partial v}{\partial\mu} = \frac{\partial P/\partial v}{\partial\mu/\partial v} = \frac{1}{v}$$

$$\frac{\partial^2 P}{\partial\mu^2} = -\frac{1}{v^2}\frac{\partial v}{\partial\mu} = \frac{\kappa_T}{v^2} \tag{9.23}$$

where

$$\kappa_T = -\frac{1}{v}\left(\frac{\partial v}{\partial P}\right)_T \tag{9.24}$$

is the isothermal compressibility. Thus we obtain

$$\frac{\overline{n^2} - \overline{n}^2}{\overline{n}^2} = \frac{k_BT\kappa_T}{V} \tag{9.25}$$

When $V \to \infty$, this vanishes unless $\kappa_T \to \infty$, which happens at the critical point.

The strong density fluctuations at the critical point occurs on a molecular scale, where the atoms come together momentarily to form large clusters, only to break

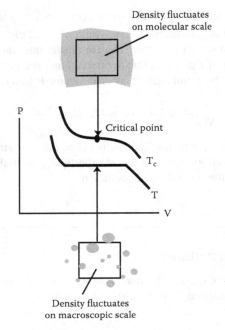

Figure 9.2   Critical opalescence at the critical point is due to the intense scattering of light by density fluctuations on a molecular scale. Density fluctuations on a macroscopic scale, such as that in a liquid-gas mixture, do not scatter light as strongly.

loose again. The scattering cross section of light is proportional to the mean-square fluctuation of density, and becomes very large at the critical point. This gives rise to the phenomenon of *critical opalescence*. In CO_2, the intensity of scattered light scattering increases a million fold at $T_c = 304$ K, $P_c = 74$ atm, and the normally transparent liquid turns milky white.

During a first-order phase transition, the pressure is independent of volume in the transition region, and one might think that $\kappa_T = \infty$; but this is not so because κ_T refers to the compressibility of a pure phase. In the transition region, the system is a mixture, and the compressibilities of the components remain finite. It is true that the density of the mixture will fluctuate, if there is no gravity to keep the two phases apart; but such fluctuations occur on a macroscopic scale, and do not lead to opalescence. This is illustrated in Figure 9.2.

9.7   Pair Creation

The grand canonical ensemble includes systems with different particle numbers, with a mean value N determined by the chemical potential. This makes sense only if N is a conserved quantity, for otherwise the chemical potential would be zero, as in the case of photons. In the everyday world, the truly conserved quantity is $N - \bar{N}$, where N and $\bar{N}$ are respectively the number of atoms and antiatoms. It appears that N is conserved only because there are no antiatoms around at room temperature.

We consider an example in which both particle and antiparticle are important. This is the case of electrons and positrons in the interior of a star, which can be pair-created and -annihilated through the reaction

$$e^+ + e^- \rightleftarrows \text{radiation} \qquad (9.26)$$

where the radiation consists of photons with a Planck distribution at a very high temperature. The reaction establishes an average value for the conserved quantum number $N_+ - N_-$, where $N_\pm$ is the number of positrons and electrons, respectively. To find the equilibrium condition, it is not necessary to know the transition rate, which determines how long it will take to establish equilibrium.

The grand partition function is given by

$$\mathcal{Q} = \sum_{N_+=0}^{\infty} \sum_{N_-=0}^{\infty} z^{N_+ - N_-} Q_{N_+} Q_{N_-} \qquad (9.27)$$

where Q_N is the partition function for a free electron or positron gas. Writing $Q_N = e^{-\beta A_N}$, and $z = e^{\beta \nu}$, we have

$$\ln \mathcal{Q} = \ln \sum_{N_+=0}^{\infty} \sum_{N_-=0}^{\infty} \exp\{-\beta[A_{N_+} + A_{N_+-} - \nu(N_+ - N_-)]\} \qquad (9.28)$$

Assuming that the fluctuations of $N_\pm$ are small, we keep only the largest term in the sum, which is determined by the conditions

$$\frac{\partial}{\partial N_+} A_{N_+} = \nu$$

$$\frac{\partial}{\partial N_-} A_{N_-} = -\nu \qquad (9.29)$$

Thus, the chemical potentials of the two gases must be equal and opposite:

$$\mu_+ + \mu_- = 0 \qquad (9.30)$$

where $\mu = \partial A_N/\partial N$. This will determine the equilibrium ratio of electrons and positrons.

Assuming $k_B T \ll mc^2$, we use the nonrelativistic limit (See Problem 8.7)

$$\mu = k_B T \ln(n\lambda^3) + mc^2 \qquad (9.31)$$

where $\lambda = \sqrt{2\pi\hbar^2/mk_B T}$ is the thermal wavelength. It is important to keep the rest energy mc^2, because we are considering reactions that convert mass into energy and vice versa. The condition for equilibrium is then

$$k_B T[\ln(n_+\lambda^3) + \ln(n_-\lambda^3)] + 2mc^2 = 0 \qquad (9.32)$$

where $n_{\pm} = N_{\pm}/V$. We can rewrite the formula as

$$n_+ n_- = \lambda^{-6} e^{-2mc^2/k_B T} \tag{9.33}$$

Assuming the initial value

$$n_- - n_+ = n_0 \tag{9.34}$$

where $n_0 > 0$, we obtain

$$n_+^2 + n_0 n_+ - \lambda^{-6} e^{-2mc^2/k_B T} = 0 \tag{9.35}$$

For $n_+/n_0 \ll 1$, the solutions are

$$\frac{n_+}{n_0} \approx \frac{1}{\lambda^6 n_0^2} e^{-2mc^2/k_B T}$$

$$\frac{n_-}{n_0} \approx 1 + \frac{1}{\lambda^6 n_0^2} e^{-2mc^2/k_B T} \tag{9.36}$$

Problems

9.1 A lattice gas consists of N_0 sites, each of which may be occupied by at most one atom. The energy of a site is ϵ if occupied, and 0 if empty. The atoms are indistinguishable.

(a) Calculate the grand partition function $Q(z, T)$ at fugacity z and temperature T.

(b) What fraction of the sites are occupied?

(c) Find the heat capacity as a function of T at fixed z.

9.2 Carbon-monoxide poisoning happens when CO replaces O_2 on Hb (hemoglobin) molecules in the blood stream. Consider a model of Hb consisting of N sites, each of which may be empty (energy 0), occupied by O_2 (energy ϵ_1), or occupied by CO (energy ϵ_2). At body temperature 37°C, the fugacities of O_2 and CO are respectively $z_1 = 10^{-5}$ and $z_2 = 10^{-7}$.

(a) Consider first the system in the absence of CO. Find ϵ_1 (in eV) such that 90% of the Hb sites are occupied by O_2.

(b) Now admit CO. Find ϵ_2 (in eV) such that 10% of the sites of occupied by O_2.

9.3 Gas molecules can adsorb on the surface of a solid at N possible adsorption sites. Each site has binding energy ϵ, and can accommodate at most one molecule, The adsorbed molecules are in equilibrium with a gas surrounding the solid. (See sketch.) We can treat the system of adsorbed molecules in a grand canonical ensemble with temperature T and chemical potential μ

(a) If there are M adsorbed molecules, what is the energy $E(M)$ of the system. What is the degeneracy $\Gamma(M)$ of the energy?

(b) Write the grand partition function of the system as a sum over M. Determine the thermal average $\bar{M}$ as the value that maximized the summand.

(c) Suppose the gas surrounding the solid has pressure P. Calculate $\bar{M}$ using the ideal-gas expression for μ.

(d) Find $\overline{M^2} - \bar{M}^2$.

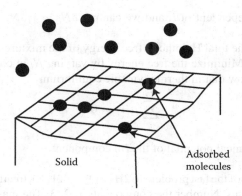

Gas of molecules

Solid

Adsorbed molecules

9.4

(a) Calculate the isothermal compressibility of a van der Waals gas near the critical volume, with $T \to T_c$ from above. Use the reduced form of the equation of state, with the critical point located at $P = V = T = 1$.

(b) Describe how the density fluctuation diverges when $T \to T_c$ from above.

9.5 In Section 9.7 we worked out the equilibrium distribution of electrons and positrons in the low-temperature limit $k_B T \ll mc^2$ Repeat the problem in the ultra-relativistic limit $k_B T \gg mc^2$. The chemical potential in this limit is given in Problem 8.7:

$$\mu \approx k_B T \ln \left(n L^3\right)$$

where

$$L = \frac{\pi^{2/3} \hbar c}{k_B T}$$

9.6 Chemical reactions A chemical reaction, such as $2H_2 + O_2 \rightleftharpoons 2H_2O$, can be denoted in the form

$$\nu_1 X_1 + \nu_2 X_2 + \cdots \rightleftharpoons \nu_1' Y_1 + \nu_2' Y_2 + \cdots$$

Taking $\nu_i' = -\nu_i$, we can rewrite this as

$$\sum_i \nu_i X_i = 0$$

The numbers ν_i are called *stoichiometric coefficients*. Consider a mixture of N_i molecules of the type X_i.

(a) Show that in a chemical reaction the changes δN_i in the numbers satisfy the relation

$$\frac{\delta N_1}{\nu_1} = \frac{\delta N_2}{\nu_2} = \cdots$$

Thus, $\delta N_i / \nu_i$ is independent of i, and we can put $\delta N_i = \nu_i \delta N$, where δN is some constant.

(b) Assume that the total Helmholtz free energy of the mixture is the sum of those of the components. Minimize the free energy by varying N_i at constant volume and temperature, and show that in thermodynamic equilibrium

$$\sum_i \mu_i \nu_i = 0$$

where μ_i is the chemical potential of the ith component.

9.7 Apply the results of the last problem to $2H_2 + O_2 \rightleftharpoons 2H_2O$, treating the components as classical ideal gases. Number the components 1, 2, 3. The masses are $m_1 = 2m$, $m_2 = 32m$, $m_3 = 18m$, where m is the nucleon mass. The stoichiometric coefficients are $\nu_1 = 2$, $\nu_2 = 1$, $\nu_3 = -2$.

(a) Show that there are two conservation laws: $n_1 - 2n_2 = A$, and $n_1 + n_3 = B$, where A, B are constants.

(b) Assume that initially there was no H_2O, that H_2 and O_2 were present in the ratio 2:1, and that the density of H_2 was n_0. Find the equation determining n_1 / n_0 as a function of temperature, and solve it in the high-temperature and low-temperature limits. Give the results for n_2 / n_0 and n_3 / n_0.

9.8 Virial coefficients The virial coefficients can be expressed in terms of the cluster integral by eliminating the fugacity z from the parametric equations of state. Treating z as a small parameter, the elimination can be done to each order in z. Show that the two lowest virial coefficients are given by

$$a_2 = -b_2$$

$$a_3 = 4b_2^2 - 2b_3$$

Chapter 10

Noise

10.1 Thermal Fluctuations

Thermodynamic quantities are supposed to be constant when the system is in thermal equilibrium. If we measure them with high precision, however, we will notice that they undergo small fluctuations. For example, the pressure a gas exerts on a wall fluctuates because of the randomness of atomic impacts. The internal energy of the gas fluctuates because it exchanges energy with the environment via atomic collisions. These fluctuations arise from the granular structure of matter, and appear as thermal noise.

We have calculated mean-square fluctuations in statistical mechanics, such as those for energy and density. We can usually ignore them, because they are vanishingly small in the thermodynamic limit. According to Equation (9.23), the mean-square fluctuation of the number of particles in a volume V is

$$\langle N^2 \rangle - \langle N \rangle^2 = \frac{N^2 k_B}{V} V \kappa_T \tag{10.1}$$

where κ_T is the isothermal compressibility. For an ideal gas this gives

$$\sqrt{\frac{\langle N^2 \rangle - \langle N \rangle^2}{\langle N \rangle^2}} = \frac{1}{\sqrt{N}} \tag{10.2}$$

Numerically, this is utterly insignificant for a macroscopic volume of gas when $N \sim 10^{23}$. However, in a volume of dimension 4000 A, of order of the wavelength of visible light, the number of atoms at STP is about 1.8×10^6, and the fractional rms fluctuation becomes 0.07%. This can be perceived indirectly through the scattering of light, as for example in the blue of the sky.

The importance of thermal noise therefore depends on the length scale of the problem. In this chapter, we study two types of noise accessible to direct observation, the Nyquist noise in an electrical resistor, and the Brownian motion of colloidal particles in suspension.

10.2   Nyquist Noise

The thermal motion of electrons in metals produces electrical noise, which is audible when amplified, as in a radio signal. The spontaneous fluctuations in the voltage $V(t)$ or current $I(t)$ average to zero, but the rms fluctuations are not zero. In circuit elements that store energy, such as a capacitor C or inductance L, these fluctuations can be obtained through the equipartition of energy: (See Problem 6.14.)

$$\frac{1}{2}C\overline{V^2} = \frac{1}{2}k_B T$$

$$\frac{1}{2}L\overline{I^2} = \frac{1}{2}k_B T \tag{10.3}$$

For a dissipative element such as a resistor, however, the fluctuation depends on its environment in a circuit.

For the spontaneous voltage fluctuation across the free ends of an open resistor, Nyquist (1928) derived the result

$$\overline{V^2} = 4Rk_B T \Delta \nu \tag{10.4}$$

where R is the resistance of the resistor, T the absolute temperature, and $\Delta \nu$ is the band width—the frequency range of the fluctuations. This result relates the voltage fluctuation to the resistance, and is an example of a fluctuation-dissipation theorem.

An intuitive argument for the result is as follows. The resistor at temperature T exchanges energy with a heat reservoir, and the average heat dissipation in the frequency range $\Delta \nu$ is

$$\overline{I^2 R} \propto k_B T \Delta \nu \tag{10.5}$$

Using Ohm's law $I = V/R$, we obtain

$$\overline{V^2} \propto Rk_B T \Delta \nu \tag{10.6}$$

For quantitative results, consider a transmission line of length L and impedance R. We terminate the transmission line at both ends with resistances R, so that traveling waves along the line are totally absorbed at the ends with no reflection. Let $V(t)$ be the voltage and $I(t)$ the current in one of the resistors. The frequency of the fundamental mode is

$$\nu_0 = \frac{c}{2L} \tag{10.7}$$

At a finite temperature T all higher modes are excited, with frequencies

$$\nu_n = n\nu_0 \quad n = 1, 2, 3, \ldots \tag{10.8}$$

The occupation number of the nth mode is $[\exp(\beta \hbar \omega_n) - 1]^{-1}$ where $\omega_n = 2\pi \nu_n$, $\beta = (k_B T)^{-1}$. Hence the energy residing in the nth mode is

$$E_n = \frac{\hbar \omega_n}{e^{\beta \hbar \omega_n} - 1} \tag{10.9}$$

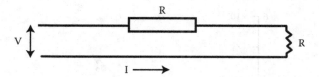

Figure 10.1 Lumped circuit diagram used in deriving the Nyquist theorem.

For $\hbar\omega_n/k_BT \ll 1$, we can use the approximation

$$E_n \approx k_BT \qquad (10.10)$$

In the band width $\Delta\nu$ there are $\Delta\nu/\nu_0$ modes, and the total energy is

$$E = \frac{k_BT\Delta\nu}{\nu_0} = \frac{2k_BTL}{c}\Delta\nu \qquad (10.11)$$

The energy can be regarded as residing in two traveling waves in opposite directions, and the time it takes to traverse the line is

$$t = \frac{L}{c} \qquad (10.12)$$

Thus the energy absorbed per second by each resistor is

$$W = \frac{E}{2t} = k_BT\Delta\nu \qquad (10.13)$$

and the power delivered to each resistor is

$$\overline{I^2}R = k_BT\Delta\nu \qquad (10.14)$$

As indicated in the lumped-circuit diagram of Figure 10.1, the voltage across a resistor is $V = 2IR$. Therefore $I = V/2R$. Multiplying both sides by I, we obtain

$$I^2R = \frac{V^2}{4R} \qquad (10.15)$$

Hence

$$\overline{V^2} = 4k_BTR\Delta\nu \qquad (10.16)$$

This is known as the *Nyquist theorem*, and predicts a universal linear relation between V^2 and R, true for all materials. Numerically the voltage fluctuation is of the order of microvolts, but Johnson (1928) measured it, and verified the Nyquist relation, as shown in Figure 10.2. He obtained Boltzmann's constant to an accuracy of 8%.

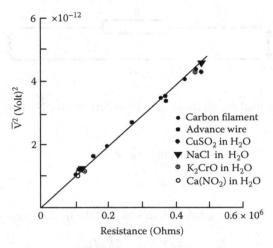

Figure 10.2 Nyquist noise: mean-square voltage fluctuation across the open ends of a resistor, as a function of resistance. The Nyquist theorem predicts the straight line, with a universal slope.

10.3 Brownian Motion

In Brownian motion, we can see with our own eyes the manifestations of molecular thermal motion. This was unsettling to the nineteenth-century mind, which would rather stick to classical mechanics and thermodynamics. However, the experiments of Perrin on this subject, which won him the physics Nobel prize in 1926, demonstrated that matter is not the pristine continuum of classical thermodynamics, but made up of noisy atoms. In his words (Perrin 1909):

> When we consider a fluid mass in equilibrium, for example some water in a glass, all the parts of the mass appear completely motionless to us. If we put into it an object of greater density it falls and, if it is spherical, it falls exactly vertically. The fall, it is true, is slower the smaller the object; but, so long as it is visible, it falls and always ends by reaching the bottom of the vessel. When at the bottom, as is well known, it does not tend again to rise, and this is one way of enunciating Garnet's principle (impossibility of perpetual motion of the second sort[1]).
>
> These familiar ideas, however, only hold good for the scale of size to which our organism is accustomed, and the simple use of the microscope suffices to impress us on new ones which substitute a kinetic for the old static conception of the fluid state.

---

[1] That is, violation of the second law of thermodynamics.

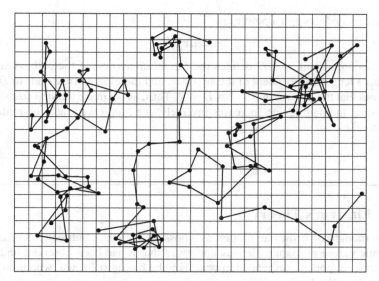

Figure 10.3 Brownian motion as sketched by Perrin, at 30-second intervals. The grid size is $3.2\,\mu$, and radius of the particle is $0.53\,\mu$.

Indeed it would be difficult to examine for long preparations in a liquid medium without observing that all the particles situated in the liquid, instead of assuming a regular movement of fall or ascent, according to their density, are, on the contrary, animated with a perfectly irregular movement. They go and come, stop, start again, mount, descend, remount again, without in the least tending toward immobility. This is the Brownian movement, so named in memory of the naturalist Brown, who described it in 1827 (very shortly after the discovery of the achromatic objective), then proved that the movement was not due to living animalculae, and recognized that the particles in suspension are agitated the more briskly the smaller they are....

The figure here reproduced (Figure 10.3) shows three drawings obtained by tracing the segments which join the consecutive positions of the same granules of mastic[2] at intervals of 30 seconds.... They only give a very feeble idea of the prodigiously entangled character of the real trajectory. If the positions were indicated from second to second, each of these rectilinear segments would be replaced by a polygonal contour of 30 sides, relatively as complicated as the drawing here reproduced, and so on.

Over a vast range of the size of time steps, the Brownian path is fractal, and a precise local velocity cannot be defined. In general, the path length L and the step

---

[2]Ingredient used in the preparation of varnish, from the bark of *Pistacia lentiscus* from Chios Island.

size τ are related through a power law:

$$L = a\tau^{1-D} \tag{10.17}$$

where a is a constant, and D defines the *fractal dimension* of the path. A smooth curve has a length independent of the step size, and therefore corresponds to $D = 1$. It turns out that the Brownian path has $D = 2$. Thus, as we make the step size smaller and smaller, the path length increases inversely, (i.e., until we reach the molecular collision time).

10.4   Einstein's Theory

Einstein thought of Brownian motion as a possible means to indirectly demonstrate the atomic nature of matter. He was not aware that it had been observed under the microscope, in pollens suspended in water, by the biologist Brown, who thought he discovered the life force.

In Einstein's theory of 1905 (Einstein 1905, 1906), he imagines that the Brownian path is divided into finite steps of equal length, and he pictures such a step, regardless of size, to be the result of a large number of smaller steps. This is the idea of a "stochastic variable," which we shall discuss more fully in the next chapter.

Consider a particle moving in one dimension, with coordinate $x(t)$. Consider the possibility that, between times t and $t + \tau$, it makes the transition

$$x \rightarrow x + \Delta \quad (-\infty < \Delta < \infty) \tag{10.18}$$

The probability for this to happen can be defined as follows. Imagine there are N such particles making the transtion, and let dN be the number of particles displaced by a distance between Δ and $\Delta + d\Delta$, during the time interval between t and $t + \tau$. The probability for the displacement to occur is the fraction of particles so displaced:

$$f_\tau(\Delta)d\Delta = \frac{dN}{N} \tag{10.19}$$

This defines $f_\tau(\Delta)$, the transition probability per unit displacement. It should be a nonnegative function with the properties

$$f_\tau(\Delta) = f_\tau(-\Delta)$$

$$\int_{-\infty}^{\infty} d\Delta \, f_\tau(\Delta) = 1 \tag{10.20}$$

The second condition means that $f_\tau(\Delta)$ should vanish faster than Δ^{-1} as $\Delta \to \infty$.

Let $n(x, t)$ be the density of particles. The number of particles being displaced out of the interval dx during time interval τ is given by

$$[n(x, t)dx] \left[\int_{-\infty}^{-dx} + \int_{dx}^{\infty} \right] d\Delta \, f_\tau(\Delta) \tag{10.21}$$

where the first factor is the number originally in dx, and the second factor is the probability of outflow. As $dx \rightarrow 0$, the gap between $-dx$ and dx in the range of integration becomes insignificant, and the outflow is 100%. Therefore, all the particles found in dx at time $t + \tau$ arrived during the time interval τ. That is,

$$n(x, t + \tau) \, dx = \int_{-\infty}^{\infty} d\Delta \, f_\tau(-\Delta) \, n(x + \Delta, t) \, dx \qquad (10.22)$$

Since $f_\tau(-\Delta) = f_\tau(\Delta)$ we have

$$n(x, t + \tau) = \int_{-\infty}^{\infty} d\Delta \, f_\tau(\Delta) \, n(x + \Delta, t) \qquad (10.23)$$

This is an expression of the conservation of particles.

We now expand $n(x, t + \tau)$ in powers of τ, and $n(x + \Delta, t)$ in powers of Δ:

$$n(x, t) + \tau \frac{\partial n(x, t)}{\partial t} + \cdots$$

$$= \int_{-\infty}^{\infty} d\Delta \, f_\tau(\Delta) \left[n(x, t) + \Delta \frac{\partial n(x, t)}{\partial x} + \frac{\Delta^2}{2} \frac{\partial^2 n(x, t)}{\partial x^2} + \cdots \right] \qquad (10.24)$$

The integral in the first term on the right is unity, and that in the second term vanishes because $f_\tau(\Delta)$ is an even function. Successive terms on the right side should rapidly become smaller, as $\tau \rightarrow 0$. Thus, we have

$$\tau \frac{\partial n(x, t)}{\partial t} = \frac{\partial^2 n(x, t)}{\partial x^2} \frac{1}{2} \int_{-\infty}^{\infty} d\Delta \, \Delta^2 f_\tau(\Delta) \qquad (10.25)$$

which becomes exact in the limit $\tau \rightarrow 0$. Assuming the existence of the limit

$$D = \lim_{\tau \rightarrow 0} \frac{1}{2\tau} \int_{-\infty}^{\infty} d\Delta \Delta^2 \, f_\tau(\Delta) \qquad (10.26)$$

we obtain the diffusion equation

$$\frac{\partial n(x, t)}{\partial t} = D \frac{\partial^2 n(x, t)}{\partial x^2} \qquad (10.27)$$

The diffusion constant D, which depends only on the second moment of the probability distribution $f_\tau(\Delta)$, can be taken from experiments.

The process described is independent of the details of the transition probability density $f_\tau(\Delta)$, as long as it decreases sufficiently fast with Δ, such that the diffusion constant D exists. A more fundamental assumption is that $f_\tau(\Delta)$ depends only on Δ, and not on previous history. Such processes are said to be Markovian, which we shall study further in Chapter 12.

10.5 Diffusion

The solution to the diffusion equation is the distribution that we have encountered several times earlier in this book:

$$n(x, t) = \frac{1}{\sqrt{4\pi Dt}} e^{-x^2/(4Dt)} \qquad (10.28)$$

with the properties

$$\int_{-\infty}^{\infty} dx\, n(x, t) = 1$$

$$n(x, t) \xrightarrow[t \to 0]{} \delta(x)$$

It gives the probability density of finding a particle at x at time t, knowing that it was at $x = 0$ at $t = 0$. The generalization of the diffusion equation to 3D is

$$\frac{\partial n(\mathbf{r}, t)}{\partial t} = D\nabla^2 n(\mathbf{r}, t) \qquad (10.29)$$

with solution

$$n(\mathbf{r}, t) = \frac{1}{(4\pi Dt)^{3/2}} e^{-r^2/(4Dt)} \qquad (10.30)$$

To verify the diffusion law experimentally, Perrin translated each sketched Brownian path parallel to itself, so that they have a common origin in the plane of the paper. The theoretical distribution is thus

$$\frac{r\,dr}{(4\pi Dt)^{3/2}} \int_{-\infty}^{\infty} dz\, e^{-(r^2+z^2)/4Dt} = \frac{2r\,dr}{\rho^2} e^{-r^2/\rho^2} \qquad (10.31)$$

where r is the radial distance in the plane, and $\rho = \sqrt{4Dt}$. For Perrin's experiments $\rho = 7.16\mu$. The distribution of 365 events is shown in Figure 10.4, and the comparison with theory is shown in Figure 10.5.

Einstein's simple and physical theory laid the foundation for a formal theory of stochastic processes. The important points are the following:

- The diffusion law is insensitive to the form of $f_\tau(\Delta)$. We can arrive at the same law by assuming that, at some small scale, a Brownian particle executes a random walk of equal step size. (See Section 5.2.)

- The Brownian displacement, which is a sum of a large number of random steps, has a Gaussian distribution. This is the *central limit theorem*.

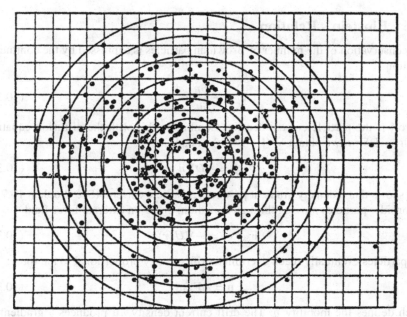

Figure 10.4 Perrin translated 365 projected Brownian paths to a common origin, in order to check the diffusion law.

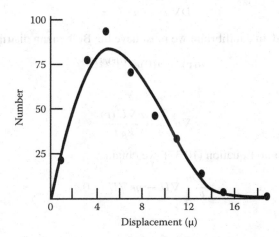

Figure 10.5 Verification of the diffusion law in Brownian motion. Solid curve is theory, and the dots represent data from Figure 10.4.

10.6   Einstein's Relation

The conservation of the number of Brownian particles is expressed by the continuity equation

$$\nabla \cdot \mathbf{j} + \frac{\partial n}{\partial t} = 0 \tag{10.32}$$

where $\mathbf{j}$ is the particle current density. Combining this with the diffusion equation $\partial n/\partial t = D\nabla^2 n$, we obtain $\nabla \cdot (\mathbf{j} + D\nabla n) = 0$, or

$$\mathbf{j} = -D\nabla n \tag{10.33}$$

This shows that a density gradient produces a particle current. (See Problem 10.7).

Consider particles moving through a medium under an external force field

$$\mathbf{F}_{\text{ext}}(\mathbf{r}) = -\nabla U(\mathbf{r}) \tag{10.34}$$

The particles eventually reach a terminal drift velocity proportional to the force:

$$\mathbf{u} = \eta \mathbf{F}_{\text{ext}} \tag{10.35}$$

which defines the mobility η. The drift current density $n\mathbf{u}$ produces a gradient in n, which in turn produces a counteracting diffusion current $\mathbf{j}$. In equilibrium these currents must balance each other, and hence

$$\mathbf{j} + n\mathbf{u} = 0 \tag{10.36}$$

or

$$-D\nabla n - \eta n \nabla U = 0 \tag{10.37}$$

On the other hand, in equilibrium we must have the Boltzmann distribution

$$n(\mathbf{r}) = n(0)e^{-U(\mathbf{r})/k_B T} \tag{10.38}$$

which gives

$$\nabla n = -\frac{n\nabla U(\mathbf{r})}{k_B T} \tag{10.39}$$

Substituting this into Equation (10.37), we obtain

$$\frac{Dn}{k_B T}\nabla U - \eta n \nabla U = 0 \tag{10.40}$$

and hence

$$D = k_B T \eta \tag{10.41}$$

This is *Einstein's relation*, historically the first fluctuation-dissipation theorem. We shall return to this at the end of the chapter.

10.7 Molecular Reality

At the end of the nineteenth century, three physical constants remained poorly known: Avogadro's number A_0, Boltzmann's constant k_B, and the fundamental charge e. Einstein's relation provides a way to experimentally measure Avogadro's number.

According to Stokes' law, a sphere moving at a terminal velocity $\mathbf{u}$ in a liquid experiences a frictional force

$$\mathbf{F} = 6\pi a v \mathbf{u} \tag{10.42}$$

where a is the radius of the sphere, and v is the coefficient of viscosity of the liquid. The mobility is therefore

$$\eta = 6\pi a v \tag{10.43}$$

By Einstein's relation we have

$$6\pi a v = \frac{D}{k_B T} \tag{10.44}$$

Using the relation $k = R/A_0$, where R is the gas constant and A_0 is Avogadro's number, we have

$$A_0 = \frac{6\pi a v R T}{D} \tag{10.45}$$

Perrin obtained Avogadro's number from this relation, among others, and arrived at a weighted average:

$$A_0 = 7.05 \times 10^{23} \quad \text{(Modern value: } 6.02 \times 10^{23}\text{)} \tag{10.46}$$

Using this, Perrin was able to obtain k_B and e. (See Problem 10.7).

Finally, we quote Perrin on molecular reality and the second law of thermodynamics:

> It is clear that this agitation (the Brownian motion) is not contradictory to the principle of the conservation of energy.... But it should be noticed, that it is not reconcilable with the rigid enunciations too frequently given to Carnot's principle[3].... One must say: "On the scale of size which interests us practically, perpetual motion of the second sort is in general so insignificant that it would be absurd to take it into account."...
>
> On the other hand, the practical importance of Carnot's principle is not attacked, and I hardly need state at length that it would be imprudent to count upon the Brownian movement to lift the stones intended for the building of a house....

---

[3]That is, the second law of thermodynamics.

I think that it will henceforth be difficult to defend by rational arguments a hostile attitude to molecular hypotheses, which, one after another, carry conviction, and to which at least as much confidence will be accorded as to the principles of energetics. As is well understood, there is no need to oppose these two great principles, the one against the other, and the union of Atomistics and Energetics will perpetuate their dual triumph.

10.8 Fluctuation and Dissipation

The atomicity of matter not only gives rise to fluctuations of thermodynamic variables, it also explains why there is friction. As Einstein's relation $D = k_B T \eta$ shows, fluctuations and dissipations are different aspects of the same physics:

- *Fluctuation*: In the absence of external force, the path of a Brownian particle fluctuates because of random molecular impacts. This is manifested through diffusion, characterized by the diffusion constant D.

- *Dissipation*: An external force is needed to drag a Brownian particle through the medium, because there is friction created by random molecular impacts, characterized by the mobility η.

These two aspects are illustrated in Figure 10.6.

The interdependence between these two aspects is quite general, and is expressed by the *fluctuation-dissipation theorem*, whose specific form depends on the system. For Brownian motion, it is derived later in Section 13.3:

$$\beta \int_0^\infty dt \langle v(t)v(0) \rangle = \eta \tag{10.47}$$

where $\beta = (k_B T)^{-1}$. On the left side we have the velocity correlation function in the absence of external field, and on the right side we have the mobility η from the

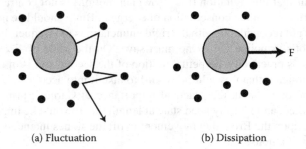

(a) Fluctuation   (b) Dissipation

Figure 10.6 Two aspects of the motion of a body in a grainy medium: fluctuation (diffusion) and dissipation (mobility). They are united through Einstein's relation.

TABLE 10.1 Various Forms of the Fluctuation-Dissipation Theorem

Quantity	Fluctuation-Dissipation Theorem	Reference
Voltage	$\beta(\overline{V^2} - \bar{V}^2) = 4R\Delta\nu$	Section 10.2
Energy	$\beta(\overline{E^2} - \bar{E}^2) = TC_V$	Section 8.5
Density	$\beta(\overline{n^2} - \bar{n}^2) = V^{-1}\bar{n}^2\kappa_T$	Section 9.5

linear response $\langle v \rangle_F = \eta F$, where $\langle v \rangle_F$ is the average velocity in the presence of the external force F. The relation for other physical systems are listed in Table 10.1, with reference to where they are discussed in this book.

It should be emphasized that the fluctuation-dissipation theorem is a linear approximation, derived under the assumptions that

- The fluctuations represent small deviations from the state of thermodynamic equilibrium.
- The system responds in a linear manner to a small disturbance.

The relation may fail in nonlinear systems, or in systems that take years to reach thermal equilibrium, such as glasses.

10.9 Brownian Motion of the Stock Market

If you look at the chart of a stock price, it's clear that there is noise; random fluctuations are ever present. Louis Bachelier, student of Henri Poincaré, was the first to model the financial market in terms of Brownian motion. In fact, he derived the diffusion equation before Einstein did (Bachelier, 1900).

If there is Brownian motion in the stock market, one would not find it in the stock price, because the price must be positive, and hence cannot have a Gaussian distribution. Instead, one finds it in the logarithm of the price (Fama 1970). Suppose the stock price is known at time $t = 0$. Let

$$p(t) = \text{stock price at time } t$$

$$s(t) = \log p(t) \tag{10.48}$$

In the Brownian model, $s(t)$ is taken to be a stochastic variable with a Gaussian distribution, whose width increases like $\sqrt{t}$, with a drifting center:

$$P(s) = \frac{1}{\sqrt{2\pi\sigma^2 t}} \exp\left(-\frac{(s - s_0 - \mu t)^2}{2\sigma^2 t}\right) \tag{10.49}$$

where $s_0 = s(0)$, and μ is the drift rate. The quantity σ is known as the *volatility*. It is of dimension $\sqrt{\text{time}}$, and commonly used as a measure of risk.

In Figure 10.7 we compare Equation (10.49) with actual data. We plot the probability distribution of $s(t) - s(0)$ for $t = 1$ day, for the SP500 index during 2003 to

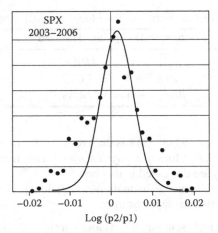

Figure 10.7   Data points represent the distribution of log (p2/p1) for SP500 during 2003–2006, with p1 = price, p2 = next day's price. Solid curve is a Gaussian fit.

2006. We can see that the distribution is roughly Gaussian, with a small drift, but the tails of the data are much fatter than Gaussian. This indicates that there are departures from Einstein's model of Brownian motion. Most probably at fault is the assumption that the transition probability does not depend on history, that is, the Markovian hypothesis. From a practical point of view, the graph shows that volatility is not a useful measure of risk for large fluctuations.

For another comparison, this time with a single stock, we consider the increment of $s(t)$ over a time interval Δt:

$$\Delta s \equiv \log \, p(t + \Delta t) - \log \, p(t) = \log \frac{p(t + \Delta t)}{p(t)} \qquad (10.50)$$

The Brownian model says

$$\Delta s = (\mu + w)\Delta t \qquad (10.51)$$

where $w \Delta t$ is a random walk. Figure 10.8 (a) shows the weekly price chart of Nutri System (NTRI) from 2003 to 2006. Panel (b) shows Δs, with $\Delta t = 1$ week. Panel (c) shows the 10-day price variance. We see that Δs does appear to be fluctuating at random. The fact that there are more points above zero than below shows a general drift. But the drift is unsteady, as the price chart shows. The variance chart shows a more or less constant band, but large fluctuations occur periodically. This indicates that the Brownian model can describe qualitative features of a "normal" market. It breaks down when there are large market moves, for reasons beyond the model.

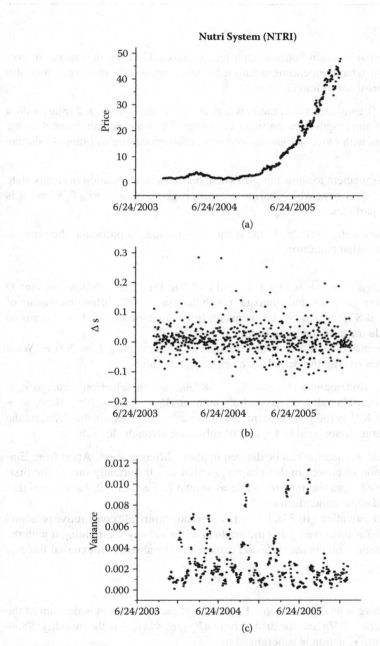

Figure 10.8 Charts of NTRI. (a) Weekly price chart. (b) Increment of $s = \log(\text{price})$. (c) Price variance over previous 10 days.

Problems

10.1 The Nyquist theorem holds at high temperatures. Find the first quantum correction to it. At what temperature would it be necessary to take this correction into account, for a resistor of length 1 mm?

10.2 Model a Brownian displacement as a path of fractal dimension 2. Start with a straight line of unit length between two fixed points. Draw a new path connecting the same endpoints with twice the length, and repeat this procedure as often as you can manage to do.

10.3 Give an arguement to show that Brownian particles in suspension in steady state behave like an ideal gas., whose partial pressure is given by $p = nk_B T$, where n is the density of particles.

10.4 Verify the solution (10.28) to the diffusion equation. In particular the normalization and the initial condition.

10.5
   (a) Perrin cited $\rho \equiv \sqrt{4Dt} = 7.16\mu$, and $t = 30$s. Find the diffusion constant D for the Brownian particles, and compare it with that for the self-diffusion constant of O_2 molecules at STP, which can be estimated from Equation (7.32). How far would an O_2 molecule travel in the same time?
   (b) Find the mobility of a Brownian particle at a temperature $T = 300$ K. What force is required to drag the particle at a velocity of 1 cm/s?

10.6 Calculate Boltzmann's constant $k_B = R/A_0$, and the electronic charge $e = F/A_0$, using Perrin's value of Avogadro's number $A_0 = 7.05 \times 10^{23}$. Here, $R = 8.32 \times 10^7$erg K^{-1} is the gas constant, and $F = 2.9 \times 10^{14}$ esu is the Faraday, the amount of charge dissociated in 1 g mol of substance through electrolysis.

10.7 The diffusion equation can be derived in many different ways. Apart from Einstein's derivation discussed in this chapter, (which was historically one of the first) it can be derived from the random walk, as shown in Section 5.2. Here is another derivation, and some generalizations.
   (a) Start with Equation (10.33), $\mathbf{j} = -D\nabla n$, as an empirical "constitutive relation" between the diffusion current $\mathbf{j}$ and the gradient of the density. Combining it with the continuity equation, obtain the diffusion equation in the absence of external force.

$$\frac{\partial n}{\partial t} = D\nabla^2 n$$

   (b) When there is an external force $\mathbf{F}_{ext}$, the total particle current is the sum of the diffusion current $-D\nabla n$ and the drift current $n\mathbf{F}_{ext}/\eta$, where η is the mobility. Show that the diffusion equation is generalized to

$$\frac{\partial n}{\partial t} = D\nabla^2 n - \eta \mathbf{F}_{ext} \cdot \nabla n$$

This is a special case of the Fokker–Planck equation.

(c) If the diffusing particle can be absorbed by the medium with an absorption probability per second $V(\mathbf{r})$, show that the diffusion equation is generalized to

$$\frac{\partial n}{\partial t} = D\nabla^2 n - Vn$$

In pure-imaginary time, this becomes the Schrödinger equation, with $D = \hbar/2m$.

References

Bachelier, L., *Theorie de la Speculation* (1900). [English translation: *The Random Character of Stock Market Prices*, H. Cootner (ed.), MIT Press, Cambridge, 1964, pp. 17–78.]

Einstein, A., *Ann. d. Phys.*, **17**:549 (1905); **19**:371 (1906).

Fama, E., *J. Finance*, **25**:383 (1970).

Johnson, J.B., *Phys. Rev.*, **32**:97 (1928).

Nyquist, H., *Phys. Rev.*, **32**:110 (1928).

Perrin, M.J., *Ann. Chem. et Phys.*, 8me series (Sep., 1909). [Translated from French by F. Soddy, reprinted in M.J. Nye, "The Question of the Atom: From the Karlsruhe Congress to the Solvay Conference, 1860–1911," Tomash Publishers, Los Angeles, 1984.]

Chapter 11

Stochastic Processes

11.1 Randomness and Probability

In physics, the term stochastic[1] refers to probabilistic considerations, and the notion of probability is based on the frequency of occurrence of random events. From a physical point of view, therefore, the central questions are

- What is meant by randomness?
- How can we decide whether a process is random?

According to our intuitive notions, a random event is one whose outcome is uncertain and unpredictable. More specifically, it is an event for which an infinitesimal change in the initial conditions will produce a very different outcome.

The flipping of a "true" coin has 50% chance of being heads or tails. That is to say, the outcome is random. Actually, this defines what we mean by a true coin. If we toss a particular coin a million times, and heads turned up 49.9% of the time, we would conclude that either the coin is biased, or we have not tossed it enough times. For us to revise our notion of a true coin will require the shock of monumental failures. From this point of view, probability is a physical concept no different from any other.

We now formulate these ideas more formally (A general reference for probability is Feller 1968.) Imagine that n independent experiments are performed under identical conditions. If outcome A is obtained in n_A of the experiments, then the probability that A occurs is

$$P_A = \lim_{n \to \infty} \frac{n_A}{n} \tag{11.1}$$

The following basic properties of probability follow from this definition:

- If two events A and B are mutually exclusively, then the probability that either A or B occurs is the sum of their probabilities $P_A + P_B$.
- If two events A and B are independent of each other, then the probability for their simultaneous occurrence is the product of their probabilities $P_A\, P_B$.

---

[1] From the *Oxford English Dictionary*: Stochastic, a. Now *rare* or *obs*. Pertaining to conjecture. 1720 Swift, *Right of Preced. betw. Physicians & Civilians 11*, I am Master of the Stochastik Art, and by Virtue of that, I divine, that those Greek Words have crept from the Margin into the Text.... 1688 Cudworth, *Freewill* (1838) 40. There is need and use of this stochastical judging and opining concerning truth and falsehood in human life.

In an experiment whose outcome must be one of a number of mutually exclusive events labeled $1, \ldots, K$, the probability P_i associated with the ith event is a real number satisfying

$$P_i \geq 0$$

$$\sum_{i=1}^{K} P_i = 1$$

There are practical difficulties in using this definition to experimentally measure probability. First, we do not have infinite time at our disposal; secondly no two actual experiments can be completely identical. Thus, the true function of this definition is to guide us in assigning *a priori* probabilities to events. Whether or not the assignment is appropriate is determined by confronting our theory with reality.

A *stochastic variable* is a quantity whose possible values occur with a certain probability distribution. It is defined when we give

- the range of its possible values y;
- the probability $P(y)$ for the occurrence of y.

The sum of two stochastic variables is a stochastic variable.

The simplest stochastic variable is the outcome y for tossing a coin. Suppose the coin is biased, so that the probability for heads is p, and that for tails $(1 - p)$. The possible values may be taken to be

$$y = \begin{cases} 1 & \text{(heads)} \\ 0 & \text{(tails)} \end{cases} \tag{11.2}$$

with probability

$$P(1) = p$$

$$P(0) = 1 - p \tag{11.3}$$

In Einstein's theory of Brownian motion discussed in the last chapter, the displacement Δ over a time interval τ is a continuous stochastic variable, with range of $-\infty < \Delta < \infty$. The probability that it has a value between Δ and $\Delta + d\Delta$ is $f_\tau(\Delta) d\Delta$. In that theory in the limit $\tau \to 0$, only the second moment of $f_\tau(\Delta)$ is relevant.

11.2   Binomial Distribution

To put our notion of stochastic variables to use, we derive some probability distributions that can serve as tests for randomness.

Consider first the sum of a number of stochastic variables, for example, the result of tossing a coin n times independently:

$$k = y_1 + \cdots + y_n \tag{11.4}$$

The possible values are

$$k = 0, 1, \ldots, n \tag{11.5}$$

The probability $P(k)$ can be obtained through the binomial theorem, which states that, in the expansion of $(1+x)^n$ in powers of x, the coefficient of x^k is the binomial coefficient

$$\binom{n}{k} = \frac{n!}{k!(n-k)!} \tag{11.6}$$

Let us write

$$(1+x)^n = \underbrace{(1+x)\cdots(1+x)}_{n \text{ factors}} \tag{11.7}$$

To obtain x^k, we pick one x each from k of the factor on the right side. The coefficient of x^k is the number of ways we can choose k factors out of the n factors. Therefore, $\binom{n}{k}$ is the number of ways to choose k things out of n things.

Now, imagine n identical coins being tossed simultaneously. There are $\binom{n}{k}$ ways of choosing the k coins that turn up heads. The probability for each of the choices is $p^k(1-p)^{n-k}$, where p is the probability for heads. The probability $P(k)$, which we redesignate as $B(k; n, p)$, is thus given by

$$B(k; n, p) = \frac{n!}{k!(n-k)!} p^k (1-p)^{n-k} \tag{11.8}$$

This is called the *binomial distribution*, the probability for getting k heads in n tosses of a coin, with p the intrinsic probability for heads. For a true coin $p = 1/2$.

The first few moments of the distribution are

$$\sum_{k=0}^{n} B(k; n, p) = 1$$

$$\langle k \rangle = \sum_{k=0}^{n} k B(k; n, p) = np$$

$$\langle k^2 \rangle = \sum_{k=0}^{n} k^2 B(k; n, p) = n^2 p^2 + np(1-p) \tag{11.9}$$

The variance, or mean-square fluctuation, is

$$\langle k^2 \rangle - \langle k \rangle^2 = np(1-p) \tag{11.10}$$

A graph of the distribution is shown in Figure 11.1.

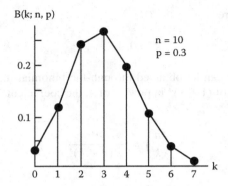

Figure 11.1 The binomial distribution [Equation (11.8)] gives the probability of getting k heads in n tosses of a biased coin, with p the intrinsic probability for heads.

11.3 Poisson Distribution

The *Poisson distribution* is the limit of the binomial distribution when $p \to 0$ and $n \to \infty$, with fixed $np = \sigma$:

$$P(k; \sigma) = \lim_{n \to \infty} B\left(k; n, \frac{\sigma}{n}\right) \tag{11.11}$$

It is known as the "law of small probabilities" for this reason. With the help of Stirling's formula, it is straightforward to obtain

$$P(k; \sigma) = \frac{\sigma^k}{k!} e^{-\sigma} \tag{11.12}$$

The first few moments are

$$\sum_{k=0}^{\infty} P(k; \sigma) = 1$$

$$\langle k \rangle = \sum_{k=0}^{\infty} k P(k; \sigma) = \sigma$$

$$\langle k^2 \rangle = \sum_{k=0}^{\infty} k^2 P(k; \sigma) = \sigma(\sigma + 1) \tag{11.13}$$

Thus the variance is

$$\langle k^2 \rangle - \langle k \rangle^2 = \sigma \tag{11.14}$$

A 3D representation of the Poisson distribution is shown in Figure 11.2.

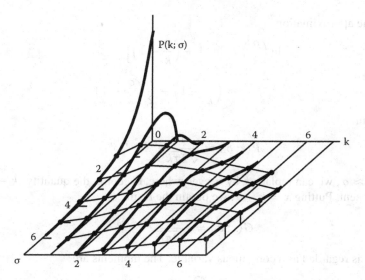

Figure 11.2 The Poisson distribution [Equation (11.12)]. The variable k is discrete, while σ is continuous. Continuous curves are indicated in heavy lines. The dots at large constant σ approach a continuous Gaussian distribution in k.

As an example of the Poisson distribution, suppose you try to sell walrus whiskers on the telephone. The demand for the product is essentially zero, but you make a thousand phone calls. What is the chance that you will make ten sales? Twenty sales? If you know the probability of a sale, you can calculate the chances via the Poisson distribution.

If you don't know the probability, and want to find out, you can make a thousand phone calls every day for a year. At the end of the period you tally up the number of days $N(k)$ in which you made $k = 0, 1, 2, \ldots$ sales. If there is enough data, the histogram of $N(k)$ should resemble a Poisson distribution. You can then normalize it so that the total area is 1, and try to fit it with a Poisson distribution by adjusting σ. The probability of a sale is then $p = \sigma/1000$.

11.4 Gaussian Distribution

The Poisson distribution gives the probability for discrete counts k. It is peaked about $k = \sigma$, with width $\sqrt{\sigma}$. For sufficiently large σ, the values of k in the neighborhood of the peak can be treated as continuous, and the Poisson distribution goes over to the *Gaussian distribution*, also known as *normal distribution*.

In the neighborhood of $k = \sigma$, for $\sigma \gg 1$, we can use the Stirling formula to write

$$P(k; \sigma) \approx \frac{1}{\sqrt{2\pi k}} e^{k - \sigma + k \ln(\sigma/k)} \tag{11.15}$$

Using the approximation

$$\ln\left(\frac{\sigma}{k}\right) = \ln\left[1 + \left(\frac{\sigma}{k} - 1\right)\right]$$

$$\approx \left(\frac{\sigma}{k} - 1\right) - \frac{1}{2}\left(\frac{\sigma}{k} - 1\right)^2 \qquad (11.16)$$

we obtain

$$P(k; \sigma) \approx \frac{1}{\sqrt{2\pi k}} e^{-(k-\sigma)^2/2k} \qquad (11.17)$$

Since $k \approx \sigma$, we can replace k by σ everywhere, except in the quantity $(k - \sigma)^2$ in the exponent. Putting $x = k - \sigma$, we obtain the result

$$G(x; \sigma) = \frac{1}{\sqrt{2\pi\sigma}} e^{-x^2/2\sigma} \qquad (11.18)$$

where x is regarded as a continuous variable. The moments are

$$\int_{-\infty}^{\infty} dx\, G(k; \sigma) = 1$$

$$\langle k \rangle = \int_{-\infty}^{\infty} dx\, x G(k; \sigma) = 0$$

$$\langle k^2 \rangle = \int_{-\infty}^{\infty} dx\, x^2 G(k; \sigma) = \sigma \qquad (11.19)$$

Clearly, the variance is σ. Graphs of the Gaussian distribution for different σ, are shown in Figure 11.3.

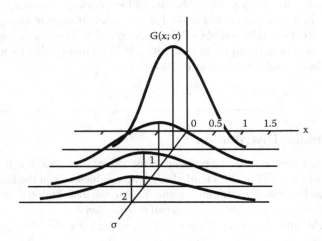

Figure 11.3 The Gaussian distribution [Equation (11.18)]. The central limit theorem states that the sum of a large number of stochastic variables obeys this distribution, independent of the probability distributions of the summands.

11.5   Central Limit Theorem

We often work with the sum of a large number of stochastic variables. For example, to test the trueness of a coin, we make a large number n of tosses, with outcomes y_i ($i = 1, \ldots, n$), and consider the probability distribution of $k = \sum_i y_i$. For sufficiently large n, we should have the Gaussian distribution [Equation (11.18)], which is centered at $\sigma = np$, with width $\sqrt{\sigma}$. Changing the variable to

$$z = \frac{1}{n} \sum_{i=1}^{n} y_i \qquad (11.20)$$

we obtain the Gaussian distribution

$$\frac{1}{\sqrt{2\pi\sigma}} e^{-(z-p)^2/2\sigma} \qquad (11.21)$$

The coin is true if $p = 1/2$.

As another example, let y_1, y_2, ... represent the displacements in the coordinate of a Brownian particle at successive time intervals $\tau \approx 10^{-10}$ s, on a molecular scale. If we make observations at 1-second intervals, we are measuring the sum

$$x = \Delta_1 + \Delta_2 + \cdots + \Delta_n \qquad (11.22)$$

where n is of the order 10^{10}. We learned in the last chapter that, regardless of the detailed form of the probability $P(\Delta)$, the distribution of x is a Gaussian

$$\frac{1}{\sqrt{4\pi Dt}} e^{-x^2/(4Dt)} \qquad (11.23)$$

where D is proportional to the second moment of $P(\Delta)$, and t is the observation time in seconds.

Summarizing these results, we can assert:

> *The sum of a large number of stochastic variables obeys the Gaussian distribution, regardless of the probability distributions of the individual stochastic variables.*

This is the *central limit theorem*. The important property of the Gaussian distribution is its universality, which makes its "bell curve" so ubiquitous, from error analysis in laboratory experiments to the forecast of longevity in a population.

11.6   Shot Noise

Shot noise consists of a series of events randomly distributed over a long period of time. Practical examples abound:

- raindrops impinging on a window pane during a rainstorm;
- electrons arriving at the anode of a vacuum tube;

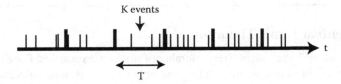

Figure 11.4   Shot noise consists of a stream of events distributed randomly over a long period of time. Here, the time axis is divided into periods of equal durations, and the collection of these periods forms a statistical ensemble.

- cars crossing an intersection during rush hour;
- customers going up to a service counter.

In reality these events are subject to modulations, of course. But, during a stretch of time when they appear to be in a steady state, we model them as a random stream.

We can map this problem into the coin-tossing problem discussed earlier. Let us divide the time axis into bins, each of duration T, and let K_i be the number of events in the ith bin, as shown in Figure 11.4. We regard the collection of bins as a statistical ensemble describing the process. The average frequency of events is defined by

$$v = \lim_{M \to \infty} \frac{K_1 + \cdots + K_M}{MT} \tag{11.24}$$

Let the unit of time be Δt. The probability of a hit during Δt is $p = v\Delta t$. We consider $\Delta t \to 0$, so p becomes vanishingly small. Then, getting a hit is like getting heads in tossing a bias coin, with intrinsic probability p. There are $n = T/\Delta t$ intervals in the bin, and we think them as n tosses of the coin. Thus, the probability of getting k hits in n tries is given by the binomial distribution $B(k; n, p)$. Now go to the limit $n \to \infty$, $p \to 0$, with fixed $np = \sigma$, given by

$$\sigma = vT \tag{11.25}$$

Then, the probability of getting k hits in time T is given by the Poisson distribution

$$P(k; \sigma) = \frac{\sigma^k}{k!} e^{-\sigma} \tag{11.26}$$

The average number of hits is $\langle k \rangle = \sigma$. The distribution depends on T only through σ.

Suppose an event happening at time $t = 0$ produces a measurable effect $f(t)$, such as the sound of a rain drop, or a current triggered by an electron. The output from the streams of random events is represented by the function

$$I(t) = \sum_{k=-\infty}^{\infty} f(t - t_k) \tag{11.27}$$

Generally, $f(t)$ makes a "plop" that is zero before $t = 0$, and is significant only for a finite time Δ. Thus, at any given t, only a finite number of events contribute

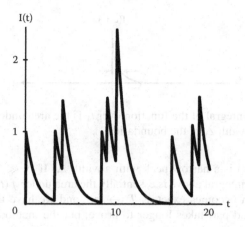

Figure 11.5 The sound of raindrops. A drop arriving at $t = 0$ produces sound represented by $f(t) = \theta(t)e^{-t}$.

significantly to the output function, namely, those that lie within the width Δ. As an illustration, let us represent the sound of a raindrop arriving at time $t = 0$ by

$$f(t) = \begin{cases} 0 & (t < 0) \\ e^{-\lambda t} & (t > 0) \end{cases} \tag{11.28}$$

where λ is a constant. Then $I(t)$ has the form shown in Figure 11.5.

Campbell's theorem states the simple results

$$\langle I(t) \rangle = \nu \int_{-\infty}^{\infty} dt\, f(t)$$

$$\langle I^2(t) \rangle - \langle I(t) \rangle^2 = \nu \int_{-\infty}^{\infty} dt\, f^2(t) \tag{11.29}$$

where $\langle I(t) \rangle$ denotes the average over an ensemble of time periods of duration T, with t held fixed in that period. The average turns out to be independent of t, in the limit $T \to \infty$.

To show the first statement of Campbell's theorem, we first consider the cases where exactly k events arrive in the period T, and average over the arrival times $t_1, \ldots, t_k$. We then average over the Poisson distribution for k:

$$\langle I(t) \rangle = \sum_k P(k; \sigma) \int_0^T \frac{dt_1}{T} \cdots \int_0^T \frac{dt_k}{T} [f(t - t_1) + \cdots f(t - t_k)]$$

$$= \sum_k k P(k; \sigma) \int_0^T \frac{dt_1}{T} f(t - t_1) = \nu \int_0^T dt_1 f(t - t_1) \tag{11.30}$$

where we have used the fact that the average of k over the Poisson distribution gives σ, and $\sigma/T = \nu$.

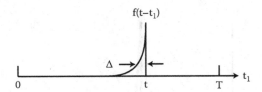

Figure 11.6 The integral of the function over t_1 is the area under the curve, except when t is within a width Δ of the boundaries.

The function $f(t)$ is a narrow peak with a width Δ. If $\Delta \ll T$, then f is like a δ-function, and the integral above is essentially the area under $f(t)$, that is, $\int dt f(t)$. This approximation becomes exact as $T \to \infty$, and we have the first part of the theorem. The second part takes longer to prove, but the method is the same. (See Problem 11.7.)

Let $P(I)dI$ be the probability that the value of $I(t)$ lies between $I + dI$ and I. For large frequency ν of the events, $I(t)$ is a sum of a large number of stochastic variables, and therefore obeys a Gaussian distribution by the central limit theorem. Thus we have

$$P(I) = \frac{1}{\sqrt{2\pi b}} e^{-(I-a)^2/2b} \tag{11.31}$$

where

$$a = \langle I(t) \rangle$$

$$b = \langle I^2(t) \rangle - \langle I(t) \rangle^2 \tag{11.32}$$

Problems

11.1 There were 12 rain showers last week, but none happened on Tuesday. Is it safe to leave the umbrella home this Tuesday?

11.2 A student drives to school and parks illegally on the street. One week she received 12 parking tickets, all on either Monday or Wednesday. Would you advise her to park in a pay lot on Mondays and Wednesdays?

11.3 A man in New York gets off work at approximately 5 PM every day, and walks to the subway station, where he could take a uptown or downtown train from the same platform. The man takes the first train that comes. (Need we explain? He has a house in Yonkers and a condo in Brooklyn.) The trains run on a strict schedule: a northbound train leaves every 5 minutes, and a southbound train leaves every 5 minutes. Over the years he wound up in Brooklyn 70% of the time. Why?

11.4 Describe a procedure to test the quality of a random-number generator, based on the Poisson distribution. Carry it out in your personal computer.

11.5 The behavior of a certain type of semiconductor diode can be modeled on the current-voltage characteristic

$$I = \begin{cases} 0 & (V < 0) \\ I_0[\exp(V/V_0) - 1] & (V \geq 0) \end{cases}$$

Find the probability density for the current in terms of the probability density for the voltage.

11.6 A device squares the input: $y = \alpha x^2$. Suppose the input x has a Rayleigh probability density

$$P(x) = \begin{cases} (x/a) \exp(-x^2/2a) & (x \geq 0) \\ 0 & (x < 0) \end{cases}$$

Find the probability density for y.

11.7 Campbell's theorem The sound of raindrops is represented by the output function $I(t) = \sum_k f(t - t_k)$, as defined in Equation (11.27). Follow the step outlined below to show that its correlation function is given by

$$G(\tau) \equiv \langle I(t)I(t + \tau) \rangle = \nu \int_{-\infty}^{\infty} dt\, f(t) f(t + \tau) + \left(\nu \int_{-\infty}^{\infty} dt\, f(t) \right)^2$$

where ν is average frequency of raindrops. The average $\langle\rangle$ denotes an average over an ensemble of time periods of duration T, with fixed t and τ, in the limit $T \to \infty$. For $\tau = 0$, this is the second part of Campbell's theorem.

(a) Consider first exactly K raindrops falling during the period T, average over the times at which they fall. Then average over a Poisson distribution $P_\sigma(K)$ of values of K. Show

$$G(\tau) = \sum_{K=0}^{\infty} P_\sigma(K) \sum_{i=1}^{K} \sum_{j=1}^{K} \int_0^T \frac{dt_1}{T} \cdots \int_0^T \frac{dt_K}{T} f(t - t_i) f(t + \tau - t_j)$$

where $\sigma = \nu T$, and the times T, t, τ, t_i are all integers in appropriate units.

(b) Consider separately the contribution from terms with $i = j$, and those with $i \neq j$. Show

$$G(\tau) = \sum_{K=0}^{\infty} P_\sigma(K) \left[\frac{K}{T} \int_0^T dt_1\, f(t - t_1) f(t + \tau - t_1) \right.$$

$$\left. + \frac{K(K-1)}{T^2} \int_0^T dt_1\, f(t - t_1) \int_0^T dt_2\, f(t + \tau - t_2) \right]$$

(c) Since $f(t)$ has a finite width, and $T \to \infty$, show that for all values of $t + \tau$ except for a negligible set within a width of the boundaries of $[0, T]$, we have

$$G(\tau) = \sum_{K=0}^{\infty} P_\sigma(K) \left\{ \frac{K}{T} \int_{-\infty}^{\infty} dt_1\, f(t) f(t + \tau) \right.$$

$$\left. + \frac{K(K-1)}{T^2} \left[\int_{-\infty}^{\infty} dt\, f(t) \right]^2 \right\}$$

Obtain the final form by summing over K.

(d) For $f(t) = \theta(t) e^{-\lambda t}$, show

$$G(\tau) = \left(\frac{\nu}{\lambda} \right)^2 + \left(\frac{\nu}{2\lambda} \right) e^{-\lambda |\tau|}$$

Reference

Feller, W., *An Introduction to Probability Theory and Its Applications*, 3rd ed., Wiley, New York, 1968.

Chapter 12

Time-Series Analysis

12.1 Ensemble of Paths

A time-series is a stochastic variable $v(t)$ that depends on time. We have seen an example of this in shot noise. Here, we study its dynamical aspects in greater depth (Wang and Uhlenbeck 1945). For concreteness we can think of $v(t)$ as the velocity of a Brownian particle, or the current in a flow. Being a stochastic variable, $v(t)$ is not a function in the usual sense, but a member of an ensemble of functions, as illustrated in Figure 12.1. The various records of $v(t)$ describe the time evolution of identically constituted systems, under the action of random forces in the environment.

At each instant of time t there is a distribution of v-values, as we can see from Figure 12.1. To describe the time series completely, however, we need more; we need to know the correlations in time. These are given through a hierarchy of probability distributions, and are listed in the following, where we use the abbreviation 1 for $\{v_1, t_1\}$:

$$W_1(1)$$

$$W_2(1, 2)$$

$$W_3(1, 2, 3) \quad \text{etc.} \tag{12.1}$$

Here, $W_k(1, \ldots, k)dv_1 \ldots dv_k$ is the joint probability of finding that v has a certain value

- between v_1 and $v_1 + dv_1$ at time t_1;
- between v_2 and $v_2 + dv_2$ at time t_2;
- between v_3 and $v_3 + dv_3$ at time t_3; etc.

Clearly, W_k must be positive-definite, and symmetric under the interchange of $\{v_i, t_i\}$ with $\{v_j, t_j\}$. The nth joint probability W_n must imply all the lower ones W_k with $k < n$:

$$W_k(1, \cdots, k) = \int_{-\infty}^{\infty} dv_{k+1} \cdots dv_n W_n(1, \ldots, n) \tag{12.2}$$

In principle, we can measure these distributions from records like those in Figure 12.1, given a sufficiently large sample. To find $W_2(1, 2)dv_1 dv_2$ from the data in Figure 12.1, for example, we follow the vertical dotted lines at t_1 and t_2, and find the fractions of records for which $v(t_1) = v_1$ and $v(t_2) = v_2$, within tolerances dv_1, dv_k, repetitively.

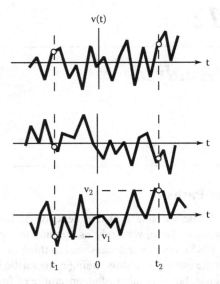

Figure 12.1   An ensemble of paths representing a stochastic process.

We shall limit our attention to *stationary ensembles*, for which the probability distributions are invariant under time translation. This means that $W_1(v_1, t_1)$ is independent of t_1, and that, for $k > 1$, W_k depends only on the relative times $(t_2 - t_1)$, $(t_3 - t_1), \ldots, (t_k - t_1)$.

Many physical processes approach a steady state after transient effects die out. In these cases, we consider only the time records taken during the steady state. In Brownian motion, for example, the velocity distribution becomes Maxwell–Boltzmann distribution after some time, (although the position never reaches equilibrium).

12.2   Ensemble Average

The ensemble average of $v(t)$ is defined as

$$\langle v \rangle = \frac{\int_{-\infty}^{\infty} dv^1 v_1 W_1(v_1, t)}{\int_{-\infty}^{\infty} dv^1 W_1(v_1, t)} = \int_{-\infty}^{\infty} dv^1 \, v_1 W_1(v_1, t) \tag{12.3}$$

The denominator in the ratio is unity by definition, because $W_1 dv_1$ is a probability. In a stationary ensemble the above is independent of the time t.

The correlation function between $v(t)$ at two different times t_1, t_2 is given by

$$\langle v(t_1)v(t_2) \rangle = \int_{-\infty}^{\infty} dv_1 \int_{-\infty}^{\infty} dv_2 \, v_1 v_2 W_1(v_1, t_1; v_2, t_2) \tag{12.4}$$

In a stationary ensemble this depends only on $t_2 - t_1$. The triple correlation function and higher ones can be similarly defined.

In a stationary ensemble, the ensemble average is equivalent to time average. For example,

$$\langle v \rangle = \frac{1}{T} \int_{-T/2}^{T/2} dt\, v(t) \tag{12.5}$$

for sufficiently large T. As an example, consider the tossing of N coins. The fraction of heads occurring should be the same, whether one coin is tossed N times in succession, or N coins are tossed simultaneously, provided N is sufficiently large. We can of course think of exceptions. It is possible that tossing the N times will result in all heads, no matter how large N may be. Such a sequence, however, is very unlikely. That is, in the space of all possible sequences it has a very small "measure" 2^{-N}. Time average and ensemble average are equivalent "except for sets of measure zero."

12.3 Power Spectrum and Correlation Function

A time series can be decomposed into its sinusoidal components through Fourier analysis:

$$v(t) = \int_{-\infty}^{\infty} \frac{d\omega}{2\pi} e^{-i\omega t} v_\omega \tag{12.6}$$

with inverse transform

$$v_\omega = \int_{-\infty}^{\infty} dt\, e^{i\omega t} v(t) \tag{12.7}$$

Since $v(t)$ is real, we must have

$$v_{-\omega} = v_\omega^* \tag{12.8}$$

The Fourier transform of the correlation function reads

$$\langle v(t_1) v(t_2) \rangle = \int_{-\infty}^{\infty} \frac{d\omega_1 d\omega_2}{(2\pi)^2} \langle v_{\omega_1} v_{\omega_2} \rangle \exp(-i\omega_1 t_1 - i\omega_2 t_2) \tag{12.9}$$

To input the information that the ensemble is stationary, let us put

$$t_1 = T + \tau/2$$

$$t_2 = T - \tau/2$$

The exponent can then be rewritten as

$$\omega_1 t_1 + \omega_2 t_2 = (\omega_1 + \omega_2)T + (\omega_1 - \omega_2)\frac{\tau}{2} \tag{12.10}$$

Thus

$$\langle v(t_1)v(t_2)\rangle = \int_{-\infty}^{\infty} \frac{d\omega_1 d\omega_2}{(2\pi)^2} \langle v_{\omega_1} v_{\omega_2}\rangle \exp\left(-i(\omega_1 + \omega_2)T - i(\omega_1 - \omega_2)\frac{\tau}{2}\right)$$

(12.11)

In a stationary ensemble, the right side should be independent of T, and that means $\langle v_{\omega_1} v_{\omega_2}\rangle$ must be zero, unless $\omega_1 + \omega_2 = 0$. That is, it must have the form

$$\langle v_{\omega_1} v_{\omega_2}\rangle = 2\pi S(\omega_1)\delta(\omega_1 + \omega_2)$$

(12.12)

The coefficient $S(\omega)$ is called the *power spectrum*, which we assume to be real, with $S(\omega) = S(-\omega)$.

The power spectrum $S(\omega)$ is a measure of the strength of various Fourier components. When $S(\omega)$ is independent of ω, all frequencies have equal weight, and we have what is called *white noise*.

Substituting Equation (12.12) into Equation (12.11), we have

$$\langle v(t_1)v(t_2)\rangle = \int_{-\infty}^{\infty} \frac{d\omega}{2\pi} S(\omega)e^{-i\omega(t_1-t_2)}$$

(12.13)

or

$$\langle v(t)v(0)\rangle = \int_{-\infty}^{\infty} \frac{d\omega}{2\pi} S(\omega)e^{-i\omega t}$$

$$= \int_0^{\infty} \frac{d\omega}{\pi} S(\omega) \cos(\omega t)$$

(12.14)

Inverting the Fourier transform gives

$$S(\omega) = \int_{-\infty}^{\infty} dt \, \langle v(t)v(0)\rangle e^{i\omega t}$$

$$= 2 \int_0^{\infty} dt \, \langle v(t)v(0)\rangle \cos(\omega t)$$

(12.15)

Thus, the power spectrum and the correlation function are Fourier transforms of each other. This relation, sometimes called the *Wiener–Kintchine theorem*, is an immediate consequence of Equation (12.12), which expresses the invariance under time translation.

Putting $t = 0$ in Equation (12.14), we have

$$\langle v^2 \rangle = \int_0^{\infty} \frac{d\omega}{\pi} S(\omega)$$

(12.16)

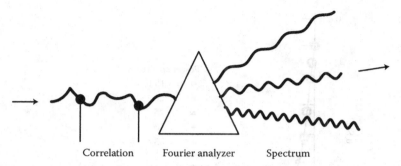

Figure 12.2   Intuitive explanation of the Wiener–Kintchine theorem, which relates spatial correlation to power spectrum: For a stationary stream of events, correlation at two different spatial points at the same time is equivalent to correlation at the same point at different times. The latter gives rise to the Fourier spectrum, while the former refers to spatial correlations.

If we think of $v(t)$ as a current, then the above represents the power dissipated in a unit resistance, and identifies $\pi^{-1} S(\omega)$ as the power dissipated per unit frequency interval.

The physical basis of the Wiener–Kintchine theorem was elucidated by G.I. Taylor (1938):

> When a prism is set up in the path of a beam of white light it analyses the time variation of electric intensity at a point into its harmonic components and separates them into a spectrum. Since the velocity of light for all wavelengths is the same, the time variation analysis is exactly equivalent to a harmonic analysis of the space variation of electric intensity along the beam.

The relation is illustrated in Figure 12.2.

Taylor also verified the Wiener–Kintchine theorem experimentally, using data on local velocities v in turbulence in a wind tunnel. The verification of Equation (12.15) requires $S(\omega)$ and $\langle v(t)v(0) \rangle$ at the same location. Equivalently one can measure $\langle v(x/u)v(0) \rangle$ for different distances x from a fixed location, where u is the average velocity of the wind. Thus, the relation to be tested can be cast in the form

$$\frac{u S(\omega)}{\langle v^2 \rangle} = \frac{2}{\langle v^2 \rangle} \int_0^\infty dx \left\langle v\left(\frac{x}{u}\right) v(0) \right\rangle \cos\left(\frac{\omega x}{u}\right) \qquad (12.17)$$

In Figure 12.3, the two sides of the above equation are obtained through independent measurements, and plotted as functions of ω/u, for a series of u's. As we can see, they are in excellent agreement.

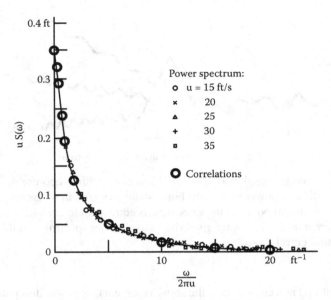

Figure 12.3   The power spectrum of turbulence in a wind tunnel, with average wind speed *u*. "Correlations" refer to the right side of Equation (12.17), which should be equal to the power spectrum by the Wiener–Kintchine theorem. [After G.I. Taylor, *op. cit.*]

12.4   Signal and Noise

From our perspective, noise is defined by the fact that all its Fourier components are stochastic variables with zero mean, that is, $\langle v_\omega \rangle = 0$. A signal, therefore, is any definite additive periodic component.

If $v(t)$ does not contain periodic components, then the correlation function approaches zero when $t \to \infty$. For the purpose of illustration, let us assume that it decays exponentially:

$$G(t) = \langle v(t)v(0) \rangle = Ce^{-\gamma|t|} \tag{12.18}$$

The corresponding power spectrum is easily found to be

$$S(\omega) = \frac{Cb}{\omega^2 + \gamma^2} \tag{12.19}$$

This is known as a *Lorentzian distribution*. Qualitative plots of $G(t)$ and $S(\omega)$ are shown in Figure 12.4.

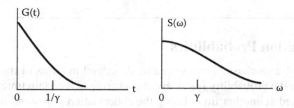

Figure 12.4 The correlation function $G(t)$ and power spectrum $S(\omega)$ for a stochastic process without periodic components. White noise corresponds to the limit $\gamma \to \infty$.

Suppose there is a periodic component, so that $v(t)$ has the form

$$v(t) = u(t) + A \sin(\omega_0 t) \tag{12.20}$$

where $u(t)$ has no periodic component. Assuming

$$\langle u(t)u(0) \rangle = C e^{-4|t|} \tag{12.21}$$

we obtain

$$G(t) = C e^{-\gamma|t|} + A^2 \sin(\omega_0 t)$$

$$S(\omega) = \frac{Cb}{\omega^2 + \gamma^2} + 2\pi A^2 \delta(\omega - \omega_0) \tag{12.22}$$

An ac signal of frequency ω_0 will show up as a spike in the power spectrum at $\omega = \omega_0$. A dc signal will give a spike at $\omega = 0$. This is illustrated in Figure 12.5.

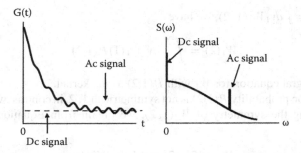

Figure 12.5 Periodic components show up as spikes in the power spectrum.

12.5 Transition Probabilities

The dynamics of a stochastic process can be described in terms of transition probabilities that give the probability for an event occurring at a certain time, when conditions are specified at another time. Using the abbreviation k to stand for $\{v_k, t_k\}$, we introduce the notation

$$P(1|2) = \text{Probability density of finding 2, when 1 is given}$$

$$P(1, 2|3) = \text{Probability density of finding 3, when 1, 2 are given}$$

$$P(1, 2, 3|4) = \text{Probability density of finding 4, when 1, 2, 3 are given}$$

$$\vdots$$

$$(12.23)$$

where probability density means probability per unit interval of the stochastic variable v. The specified variables are listed to the left of the vertical bar, and the variable to be found is written to the right. With these transition probabilities, we can express W_k in terms of $W_1, \ldots, W_{k-1}$, as follows:

$$W_2(1, 2) = W_1(1)P(1|2)$$

$$W_3(1, 2, 3) = W_2(1, 2)P(1, 2|3)$$

$$W_3(1, 2, 3, 4) = W_2(1, 2, 3)P(1, 2, 3|4)$$

$$\vdots$$

$$(12.24)$$

The transition probabilities must be positive-definite, and satisfy the normalization condition

$$\int_{-\infty}^{\infty} dv_2 P(1|2) = 1 \qquad (12.25)$$

Since $W_1(2) = \int dv_1 W_2(1, 2)$, we have

$$W_1(2) = \int_{-\infty}^{\infty} dv_1 W_1(1)P(1|2) \qquad (12.26)$$

This is an integral equation for W_1 with $P(1|2)$ as the kernel.

The transition probability $P(1|2)$ is not symmetric in 1, 2, in contrast with $W_2(1, 2)$. However, using the symmetry of $W_2(1, 2)$, we obtain from Equation (12.24) the relation

$$W_1(1)P(1|2) = W_1(2)P(2|1) \qquad (12.27)$$

That is, the transition probability weighted by the probability of the initial state is symmetric. This property is known as *detailed balance*.

12.6 Markov Process

The simplest stochastic process is a "purely random" process, in which there are no correlations. All distributions are then determined by W_1:

$$W_2(1, 2) = W_1(1)W_1(2)$$

$$W_3(1, 2, 3) = W_1(1)W_1(2)W_1(3) \quad \text{etc.} \tag{12.28}$$

Such would be the case in the successive tossing of a coin. A continuous physical variable $v(t)$ cannot be purely random, however, because $v(t)$ and $v(t + dt)$ must be correlated, for sufficiently small dt.

Next in complexity comes the *Markov process*, in which the system has no memory beyond the last transition. It is defined by the property

$$P(1, 2, \dots, n-1|n) = P(n-1|n) \qquad (t_n > t_{n-1} > \cdots > t_1) \tag{12.29}$$

All information about the process is therefore contained in $P(1|2)$ or equivalently $W_2(1, 2)$. For example, from Equation (12.24) we obtain

$$W_3(1, 2, 3) = W_2(1, 2)P(12|3) = W_2(1, 2)P(2|3)$$

$$= W_1(1)P(1|2)P(2|3) \tag{12.30}$$

The basic transition probability $P(1|2)$ cannot be arbitrary, but must satisfy the condition

$$P(1|3) = \int_{-\infty}^{\infty} dv_2 P(1|2)P(2|3) \tag{12.31}$$

More explicitly,

$$P(v_1, t_1|v_3, t_3) = \int_{-\infty}^{\infty} dv_2 P(v_1, t_1|v_2, t_2)P(v_2, t_2|v_3, t_3) \tag{12.32}$$

This is called the *Smoluchowski equation,* or *Chapman–Kolmogorov equation.* This follows from the law of composition of probabilities implied by Equation (12.2), given that $P(i|j)$ are the only independent transition probabilities.

Markov processes are important, because most physical processes are of this type. Brownian motion falls into this category, and so do quantum mechanical transitions, where the transition probability per unit time is given by "Fermi's golden rule"

$$\frac{\partial}{\partial t} P(1|2) = \frac{2\pi}{\hbar} |H'_{12}|^2 \rho_2 \tag{12.33}$$

where $t = t_1 - t_2$, H'_{12} is the matrix element of the interaction Hamiltonian, and ρ_2 is the density of final states.

12.7   Fokker–Planck Equation

We now derive an equation for $P(1|2)$ in a Markov process, under the assumption that small displacements occur over small time intervals. First, let us make explicit the invariance under time translation by writing:

$$P(1|2) = P(v_1, t_1|v_2, t_2) = P(v_1|v_2, t_2 - t_1) \tag{12.34}$$

As a suggestive notation, let us designate a transition $v \to u$ over the time t as

$$\left[v \xrightarrow[t]{} u \right] \tag{12.35}$$

If we change the time from t to $t + \Delta t$, the transition can be described as a two-step process:

$$\left[v \xrightarrow[t+\Delta t]{} u \right] = \left[v \xrightarrow[t]{} w \right] \cdot \left[w \xrightarrow[\Delta t]{} u \right] \tag{12.36}$$

The corresponding Smulochowski equation is

$$P(v|u, t + \Delta t) = \int_{-\infty}^{\infty} dw P(v|w, t) P(w|u, \Delta t) \tag{12.37}$$

We now introduce the main assumption, namely that the final state is close to the initial state, over a small time interval. This means that the transition probability over a small time Δt falls off rapidly when the final state deviates from the initial state. More precisely, consider the first two moments of the change in v:

$$\langle \Delta v \rangle \equiv \int_{-\infty}^{\infty} dw (w - v) P(v|w, \Delta t) = \int_{-\infty}^{\infty} dw\, w\, P(v|v + w, \Delta t)$$

$$\langle (\Delta v)^2 \rangle \equiv \int_{-\infty}^{\infty} dw (w - v)^2 P(v|w, \Delta t) = \int_{-\infty}^{\infty} dw\, w^2 P(v|v + w, \Delta t) \tag{12.38}$$

The assumption is that these are of order Δt, and hence the the following limits exist:

$$A(v) = \lim_{\Delta t \to 0} \frac{1}{\Delta t} \langle \Delta v \rangle$$

$$B(v) = \lim_{\Delta t \to 0} \frac{1}{\Delta t} \langle (\Delta v)^2 \rangle \tag{12.39}$$

Higher moments are assumed to be of order $(\Delta t)^2$.

Now consider the folding of $\partial P / \partial t$ with an arbitrary function R:

$$\int_{-\infty}^{\infty} du R(u) \frac{\partial P(v|u, t)}{\partial t} = \lim_{\Delta t \to 0} \frac{1}{\Delta t} \int_{-\infty}^{\infty} du R(u) [P(v|u, t + \Delta t) - P(v|u, t)] \tag{12.40}$$

We can use the Smoluchowski equation to rewrite the first term on the right side as

$$\int_{-\infty}^{\infty} du \int_{-\infty}^{\infty} dw\, R(u) P(v|w,t) P(w|u, \Delta t) \tag{12.41}$$

Interchange the order of integration, and expand $R(u)$ about $u = w$ in a Taylor series, we have:

$$R(u) = R(w) + (u - w) R'(w) + \frac{1}{2}(u - w)^2 R''(w) + \cdots \tag{12.42}$$

Substituting this into Equation (12.40) gives

$$\int_{-\infty}^{\infty} du\, R(u) \frac{\partial P(v|u,t)}{\partial t} = \int_{-\infty}^{\infty} dw\, P(v|w,t) \left[R'(w) A(w) + \frac{1}{2} R''(w) B(w) \right] \tag{12.43}$$

On the right side, we make partial integrations, change the integration variable from w to u, and rewrite it in the form

$$\int_{-\infty}^{\infty} du \left[-R(u) \frac{\partial}{\partial u}(AP) + \frac{1}{2} R(u) \frac{\partial^2}{\partial u^2}(BP) \right] \tag{12.44}$$

where $A = A(u)$, $B = B(u)$, $P = P(v|u,t)$. Thus,

$$\int_{-\infty}^{\infty} du\, R(u) \left[\frac{\partial}{\partial t} P + \frac{\partial}{\partial u}(AP) - \frac{1}{2} \frac{\partial^2}{\partial u^2}(BP) \right] = 0 \tag{12.45}$$

Since R is arbitrary, the quantity in brackets must vanish, leading to the *Fokker–Planck equation*

$$\frac{\partial P}{\partial t} + \frac{\partial}{\partial u}(AP) - \frac{1}{2} \frac{\partial^2}{\partial u^2}(BP) = 0 \tag{12.46}$$

Here, $P = P(v|u,t)$ is a function of u and t, with v as initial condition. The functions $A(u)$ and $B(u)$ specify the system under consideration. In the next chapter, we shall illustrate how they can be calculated from the dynamics of the system.

12.8 The Monte Carlo Method

The Monte Carlo method is a computer algorithm to generate a thermal ensemble. An ideal statistical ensemble consists of an infinite number of copies, and no computer can produce that. What we can do is to generate members of the ensemble one at a time. After we make the desired measurements and store the results, the member can be overwritten. Thus, the size of the ensemble would be limited by computing time instead of computer memory.

Consider a system whose state is denoted by C, and the energy of the state by $E(C)$. In the canonical ensemble with temperature T, with $\beta = 1/k_B T$, the probability for the occurrence of C in the ensemble is $e^{-\beta E(C)}$, and the thermodynamic average of any quantity $O(C)$ is given by

$$\langle O \rangle = \frac{\sum_C e^{-\beta E(C)} O(C)}{\sum_C e^{-\beta E(C)}} \qquad (12.47)$$

Our object is to instruct the computer to generate a sequence of states with the canonical distribution, that is, states should be output with relative probability $e^{-\beta E(C)}$.

Let $f(C)$ be the probability distribution of a given ensemble. The equilibrium ensemble corresponds to

$$f_{eq}(C) = \frac{e^{-\beta E(C)}}{\sum_C e^{-\beta E(C)}} \qquad (12.48)$$

We want to generate a sequence of states $C_1 \to C_2 \to \cdots C_n + C_{n+1} \to \cdots$, which starts with an arbitrary initial state C_1, and, after a "warm up" period of n steps, reaches a steady sequence of equilibrium states. From the nth step on, the "time average" with respect to the sequence should be equivalent to an ensemble average over a canonical ensemble.

The objective is achieved through a Markov process with transition probability $P(C_1|C_2)$ for $C_1 \to C_2$. It is the conditioned probability of finding the system in C_2, when it is in C_1. We impose the following conditions:

$$P(C_1|C_2) \geq 0$$

$$\sum_{C_2} P(C_1|C_2) = 1$$

$$e^{-\beta E(C_1)} P(C_1|C_2) = e^{-\beta E(C_2)} P(C_2|C_1) \qquad (12.49)$$

The first two are necessary properties of any probability. The last is the statement of detailed balance, when the system is in contact with a heat reservoir.

Theorem 12.1   *The Markov process defined by Equation (12.49) eventually leads to the equilibrium ensemble.*

Proof: Summing the detailed balance statement over states C_1, we obtain

$$\sum_{C_1} e^{-\beta E(C_1)} P(C_1|C_2) = \sum_{C_1} e^{-\beta E(C_2)} P(C_2|C_1) \qquad (12.50)$$

On the right side we note $\sum_{C_1} P(C_2|C_1) = 1$. Thus

$$\sum_{C_1} e^{-\beta E(C_1)} P(C_1|C_2) = e^{-\beta E(C_2)} \qquad (12.51)$$

This shows that the equilibrium distribution is an eigenstate of the transition matrix. The "distance" between two ensembles $f_1(C)$ and $f_2(C)$ may be measured by

$$d(f_1, f_2) = \sum_C |f_1(C) - f_2(C)| \tag{12.52}$$

Suppose f_2 is obtained from f_1 through a transition: $f_2(C) = \sum_{C'} f_1(C')P(C'|C)$. The distance between f_2 and the equilibrium ensemble is given by

$$\sum_C |f_2(C) - f_{eq}(C)| = \sum_C \left| \sum_{C'} f_1(C')P(C'|C) - f_{eq}(C) \right|$$

$$= \sum_C \left| \sum_{C'} [f_1(C') - f_{eq}(C')]P(C'|C) \right|$$

$$\le \sum_C \sum_{C'} |f_1(C') - f_{eq}(C')|P(C'|C) \tag{12.53}$$

where we have used Equation (12.51) in the second step. Putting $\sum_C P(C'|C) = 1$ in the last step, we obtain the inequality

$$\sum_C |f_2(C) - f_{eq}(C)| \le \sum_C [f_1(C) - f_{eq}(C)] \tag{12.54}$$

This shows that the distance from the equilibrium cannot decrease as the result of a transition.

The *Metropolis algorithm* gives a recipe in conformity with the rules [Equation (12.49)], as follows:

- Suppose the state is C.
- Make a trial change to C'.
- If $H(C') < H(C)$, accept the change.
- If $H(C') > H(C)$, accept the change conditionally, with probability $e^{-\beta[H(C')-H(C)]}$.

The conditional change in the last statement simulates thermal fluctuations. The relative transition probability corresponding to this algorithm is

$$T(C|C') = \begin{cases} 1 & \text{if } E(C') < E(C) \\ e^{-\beta[E(C')-E(C)]} & \text{if } E(C') > E(C) \end{cases} \tag{12.55}$$

The transition probability is obtained by properly normalizing the above:

$$P(C|C') = \frac{T(C|C')}{\sum_{C''} T(C|C'')} \tag{12.56}$$

12.9   Simulation of the Ising Model

We illustrate the Monte Carlo method with the 2D Ising model, which is a model of ferromagnetism defined on a lattice. Attached to each lattice site is a spin that can point "up" or "down," and the spins interact with nearest-neighbor interactions. Possible lattices include the square lattice and the triangular lattice, as illustrated in Figure 12.6. We shall consider the square lattice here.

Denote the spin variable by $s_i = \pm 1$. A state is specified by all the spins:

$$C = \{s_1, \cdots s_N\} \tag{12.57}$$

The index $i = 1, \ldots, N$ labels the lattice sites. Each nearest-neighbor pair has an energy $-\epsilon$ if the spins are antiparallel, and ϵ if they are parallel, where $\epsilon > 0$ for ferromagnetism, and $\epsilon < 0$ for antiferromagnetism. The energy of the system is then

$$E(C) = -\epsilon \sum_{\langle ij \rangle} s_i s_j - h \sum_i s_i \tag{12.58}$$

where $\langle ij \rangle$ denotes a nearest-neighbor pair of sites, and h is an external magnetic field. For the ferromagnetic case on a square lattice, the model is exactly soluble (Huang 1987).

The total magnetization is defined by

$$M(C) = \sum_i s_i \tag{12.59}$$

In the absence of external field, $h = 0$, a nonzero value of the ensemble average $\langle M \rangle$ indicates spontaneous magnetization. This occurs as a phase transition below a critical temperature T_c. We can numerically compute $\langle M \rangle$ by averaging $M(C)$ over a canonical ensemble, generated via the Monte Carlo method.

Set up the lattice in the computer by assigning memory locations to the sites, and choose a definite boundary condition. It is simplest to choose periodic boundary

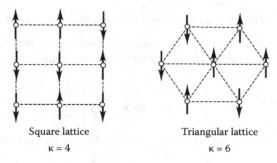

Square lattice   Triangular lattice
$\kappa = 4$ $\kappa = 6$

Figure 12.6   2D Ising model: A two-valued spin is placed on each lattice site, with nearest-neighbor interactions. The geometry of the lattice is characterized by the number of nearest neighbors κ, and their connectivity.

conditions. After initializing the lattice spins, say all spins up for a "cold" start, or random assignments for a "hot" start, we bring the lattice to thermal equilibrium by a sequence of updates. The lattice is updated by going through the spins one by one, and deciding whether or not to flip it. This is called "one sweep" of the lattice. When examining s_i, we only need to know its four nearest-neighbor spins. The interaction energy of s_i is given by

$$w_i = -\epsilon s_i \sum_{\text{nn } i} s_j - h s_i \qquad (12.60)$$

where nn stands for "nearest-neighbor to." If we flip s_i, then w_i changes sign. The Metropolis algorithm says:

$$\begin{aligned} &\text{if } w_i > 0, &&\text{flip } s_i. \\ &\text{if } w_i < 0, &&\text{flip } s_i \text{ with probability } e^{-2\beta w_i} \end{aligned} \qquad (12.61)$$

After this is done for a particular spin, we go to the next spin and repeat the algorithm, until we have gone through all the spins on the lattice. This completes one sweep of the lattice, and produces a transition $C \to C'$ We keep making sweeps, until we think the lattice is "warmed up." Then subsequent updates will generate the canonical ensemble, and we can start making "measurements."

There is a more efficient algorithm called the *heat bath method*, which is appropriate for the Ising model, or any model in which s has relatively small number of possible values. It amounts to "touching" the spins with a heat bath and thermalizing them, one at a time. The interaction energy of a spin s is

$$w = s(\Sigma + h) \qquad (12.62)$$

where Σ is the sum of nearest neighbor spins. Its possible values are

$$\Sigma = 4, 2, 0, -2, -4 \qquad (12.63)$$

We take the probability of flipping to be

$$\frac{e^{\beta w}}{e^{\beta w} + e^{-\beta w}} \qquad (12.64)$$

It is easily seen that this is a probability that satisfies detailed balance. For $s = 1$ the flip probability is

$$P = \frac{1}{1 + e^{-2\beta(\Sigma + h)}} \qquad (12.65)$$

For $s = -1$, the flip probability is

$$1 - P = \frac{1}{1 + e^{2\beta(\Sigma + h)}} \qquad (12.66)$$

We can prepare a lookup table for P, for each of the possible values of Σ. This need to be done only once to initialize the program. During runtime, we decide whether or not to flip by looking up P if $s = 1$, and $1 - P$ if $s = -1$. It can be shown that one step in the heat bath method is equivalent to an infinite number of Metropolis steps. The heat bath method becomes unwieldy if s has a large number of possible values.

Having discussed ways to compute the transition probability, we now give an overview of the computer program. We generate a sequence of states, and regard the first N_0 states as "warm-ups." The states after this are supposed to have thermalized, and should have a canonical distribution. That is, the sequence from this point on generates members of the equilibrium ensemble. The process is schematically depicted in the following:

$$\underbrace{(C_1' \to \cdots \to C_K')}_{\text{Warm-ups}} \to \underbrace{(C_1 \to C_2 \to \cdots)}_{\text{The ensemble}} \tag{12.67}$$

A member of the ensemble C_i is kept in memory only for as long as needed to perform measurements, and is overwritten by the next member. In the measurement process, we calculate the energy $E(C_i)$, the energy square $E^2(C_i)$, and the magnetization $M(C_i)$. These quantities are accumulated additively when C_i is replaced by C_{i+1}. As the program unfolds, we keep the following running totals:

$$E = E(C_1) + E(C_2) + \cdots$$

$$E^2 = E^2(C_1) + E^2(C_2) + \cdots$$

$$M = M(C_1) + M(C_2) + \cdots \tag{12.68}$$

Other quantities can be calculated and accumulated in the same fashion.

Suppose at the end of a run we have generated K states in the ensemble. We then calculate the following ensemble averages

$$\langle E \rangle = \frac{1}{K} \sum_{i=1}^{K} E(C_i)$$

$$\langle E^2 \rangle = \frac{1}{K} \sum_{i=1}^{K} E^2(C_i)$$

$$\langle M \rangle = \frac{1}{K} \sum_{i=1}^{K} M(C_i) \tag{12.69}$$

The heat capacity is given by

$$C = \frac{1}{k_B T^2} (\langle E^2 \rangle - \langle E \rangle^2) \tag{12.70}$$

These numbers constitute the output of the program in one run.

To calculate the statistical errors in the output, we have to make a large number of independent runs. The results for a measured quantity should have a Gaussian distribution, from which we can find the mean and variance.

Problems

12.1 The correlation function for "the sound of raindrops" is given in Problem 11.7.
   (a) Show that the power spectrum is given by

$$S(\omega) = u|f_\omega|^2 + \langle I(t) \rangle^2 2\pi \delta(\omega)$$

where f_ω is the Fourier transform of $f(t)$, the sound of a single raindrop, and $I = \sum_k f(t - t_k)$ is the output stream.
   (b) Obtain $S(\omega)$ for $f(t) = \theta(t)e^{-\lambda t}$ and interpret the result. Is there a white-noise component?

12.2 The shot noise in a diode is described by the voltage $V(t) = R \sum_k \varphi(t - t_k)$, where $\varphi(t)$ represents a pulse of current. The arrival times t_k are random, with an average rate v. The output is put through a low-pass filter, so only low frequencies are observed. Find the power spectrum in the low-frequency limit.

12.3 Random telegraph signals: A stream of random telegraph signals $I(t)$ in a time interval $[0, T]$ is illustrated in the accompanying figure, where $T \to \infty$ eventually. The signals have values either a or $-a$, and are of random length. The zeros on the time axis are distributed according to a Poisson distribution, with average rate v. We consider an ensemble of time intervals. Find the correlation function and power spectrum, by following the following steps.

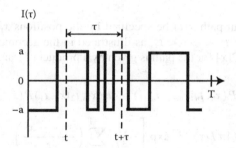

   (a) The correlation function $\langle I(t)I(t + \tau) \rangle$ is obtained by averaging the product over the ensemble, with fixed t and τ. It is independent of t because the ensemble is stationary. Referring to the illustration, we see that the product is a^2 if the factors have the same sign, and $-a^2$ if they have opposite signs. Thus show

$$\langle I(t)I(t + \tau) \rangle = a^2 P_{\text{even}} - a^2 P_{\text{odd}}$$

where P_{even}, P_{odd} are respectively the probability that there are an even and odd number of sign changes in the time interval τ.
   (b) Show

$$\langle I(t)I(t + \tau) \rangle = a^2 e^{-2v|\tau|}$$

   (c) Find the power spectrum.

12.4 Show that the Smoluchowski equation [Equation (12.31)] follows from the basic property of probabilities [Equation (12.2)] and the definitions of transition probabilities [Equation (12.24)].

Hint: Start with the equation $W_3(3, 1, 2) = \int dx_4 W_4(3, 1, 4, 2)$ from Equation (12.2), with the variables in the order shown. Express W_3 and W_4 in terms of $P(i|j)$ with the help of Equations (12.24) and (12.29).

12.5 Write and run a computer program, in your favorite language, to calculate the transition temperature T_c of the 2D Ising model using the Monte Carlo method. The exact value of T_c is given by $\tanh^2(2\epsilon/k_B T_c) = 1/2$, or

$$\frac{k_B T_c}{\epsilon} = 2.269185$$

12.6 Probability of a path If we consider an ensemble of paths $x(t)$, we should be able to assign a probability $\mathcal{P}[x]$ that is a functional of the path. We do this in this problem for the case of Brownian motion.

(a) The transition probability density of finding x at time t, given that it had the value x_0 at time t_0, is given by the following generalization of Equation (10.28):

$$P(x, t|x_0, t_0) = \frac{1}{\sqrt{4\pi D(t - t_0)}} \exp\left(-\frac{(x - x_0)^2}{4D(t - t_0)}\right)$$

(b) Let a Brownian path $x(t)$ be specified by the positions x_i at time t_i ($i = 0$, $1, \ldots, n$), with $t_0 < t_1 < \cdots < t_n$, as illustrated in the accompanying figure. The probability density $\mathcal{P}[x]$ for the path is given by a product of transition probabilities:

$$\mathcal{P}[x] = P(x_n, t_n|x_{n-1}, t_{n-1}) \cdots P(x_2, t_2|x_1, t_1) P(x_1 t_1|x_0, t_0)$$

$$= (4\pi D\tau)^{-n/2} \exp\left[-\frac{1}{4D} \sum_{i=0}^{n-1} \left(\frac{x_{i+1} - x_i}{\tau}\right)^2 \tau\right]$$

When multiplied by $dx_0 dx_1 \cdots dx_n$, this gives the probability that the Brownian particle is found between $x_i + dx_i$ and x_i, at time t_i, for $i = 0, 1, \ldots, n$. Show that it satisfies the law of composition of probabilities, that is, if we sum this probability density over all possible paths having the same endpoints, we obtain the transition probability density to go from the initial point to the final point. That is,

$$P(x_n, t_n|x_0, t_0) = \int_{-\infty}^{\infty} dx_{n-1} \cdots \int_{-\infty}^{\infty} dx_1 \mathcal{P}[x]$$

To show this, first prove the relation for $n = 1$ and $n = 2$ by direct integration. Then prove the general result by induction.

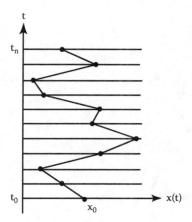

12.7 Feynman path-integral representation of diffusion

(a) In the limit $n \to \infty, \tau \to 0$, we can write

$$\mathcal{P}[x] = \mathcal{N} \exp\left(-\frac{1}{4D} \int_{t_0}^{t} dt' \dot{x}^2\right)$$

where $x(t)$ is a path, and $\dot{x}(t) = dx/dt$.

(b) The composition law becomes

$$P(x, t | x_0, t_0) = \mathcal{N} \int (dx) \exp\left(-\frac{1}{4D} \int_{t_0}^{t} dt' \dot{x}^2\right)$$

where the integral extends over all paths $x(t)$ with fixed endpoints $x(t_0) = x_0, x(t) = x_1$. This is a Feynman path integral, or "integration over all histories," as Feynman calls it. This result gives a path-integral representation of the solution to the diffusion equation. When the time is continued to pure-imaginary values, this is a solution of the free-particle Schrödinger equation in quantum mechanics.

References

Huang, K., *Statistical Mechanics*, 2nd ed., Wiley, New York, 1987, Chapter 15.

Taylor, G.I., *Proc. Roy. Soc.*, **A164** (1938). [Included in *The Scientific Papers of Sir Geoffrey Ingram Taylor*, Vol. II, G.K. Batchelor (ed.), Cambridge University Press, Cambridge, England, 1960.]

Wang, M.C. and G.E. Uhlenbeck, *Rev. Mod. Phys.*, **17**:113 (1945). [This paper, as well as other relevant ones, is included in *Selected Papers on Noise and Stochastic Processes*, N. Wax (ed.), Dover Publications, New York, 1954.]

Chapter 13

The Langevin Equation

13.1 The Equation and Solution

The Langevin equation is a simple model of the dynamics of Brownian motion. For a free Brownian particle in suspension, the Langevin equation is a special form of Newton's equation of motion:

$$m\frac{dv}{dt} + m\gamma v = F \tag{13.1}$$

where v is the instantaneous velocity and m is the mass. The total force that the medium is exerting on the particle is split into two parts:

- Random force $F(t)$.

- Frictional force $-m\gamma v$

These forces represent different aspects of interaction with the medium, one representing fluctuation, the other dissipation. They are not independent of each other, but related via the fluctuation-dissipation theorem.

The current I in a RL circuit also satisfies a Langevin equation of the form

$$L\frac{dI}{dt} + RI = V \tag{13.2}$$

where L is the inductance, R the residence, and V the voltage, which fluctuates due to Nyquist noise.

The friction coefficient γ is related to the mobility η, as we can see from the following consideration. Assume, for the present argument, that F contains a steady component F_0. In steady state, when v reaches the terminal velocity v_0, we should have $dv/dt = 0$. Thus $v_0 = (m\gamma)^{-1} F_0$, and the mobility is given by

$$\eta = \frac{1}{m\gamma} \tag{13.3}$$

In the absence of a steady component, the random force F is a stochastic variable described by an ensemble of values. The ensemble is defined through the time-correlation properties

$$\langle F(t) \rangle = 0$$

$$\langle F(t_1) F(t_2) \rangle = c_0 \delta(t_1 - t_2) \tag{13.4}$$

where $\langle\,\rangle$ denotes ensemble average. This describes white noise, for which the correlation time is zero. Physically, we are assuming that the correlation time is much shorter than the characteristic time of the problem. That is, we assume that we observe Brownian motion on a time scale much larger than the molecular collision time, which is typically 10^{-20} seconds. Thus, the velocity $v(t)$ is to be thought of as an average over many collisions. It is therefore a sum of a large number of stochastic variables, and should have a Gaussian distribution according to the central limit theorem. This will be shown later, with the help of the Fokker–Planck equation.

The Langevin equation can solved by using Fourier transforms. We write

$$v(t) = \int_{-\infty}^{\infty} \frac{d\omega}{2\pi} v_\omega \, e^{-i\omega t}$$

$$F(t) = \int_{-\infty}^{\infty} \frac{d\omega}{2\pi} F_\omega \, e^{-i\omega t} \tag{13.5}$$

with the inverse formulas

$$v_\omega = \int_{-\infty}^{\infty} dt \, v(t) \, e^{i\omega t}$$

$$F_\omega = \int_{-\infty}^{\infty} dt \, F(t) \, e^{i\omega t} \tag{13.6}$$

The Fourier transforms have the properties $v_{-\omega} = v_\omega^*$, and $F_{-\omega} = F_\omega^*$, because the quantities being transformed are real numbers.

In terms of Fourier transforms, the defining properties of the force read

$$\langle F_\omega \rangle = 0$$

$$\langle F_\omega F_{\omega'} \rangle = 2\pi c_0 \delta(\omega + \omega') \tag{13.7}$$

The transformed Langevin equation is

$$-im\omega v_\omega + \gamma v_\omega = F_\omega \tag{13.8}$$

from which we immediately obtain the solution

$$v_\omega = \frac{F_\omega}{m} \frac{1}{\gamma - i\omega} \tag{13.9}$$

This is a stochastic variable to be used in calculating correlation functions. The inverse transform of v_ω gives a particular solution $v(t)$. To get the most general solution, one must add an arbitrary homogeneous solution, that is, the solution to Equation (13.1) with $F = 0$.

For illustration, we calculate the velocity correlation functions. First, $\langle v_\omega \rangle = 0$, since $\langle F_\omega \rangle = 0$. Next we have

$$\langle v_\omega^* v_{\omega'} \rangle = \frac{1}{m^2} \frac{\langle F_\omega^* F_{\omega'} \rangle}{(\gamma + i\omega)(\gamma - i\omega')} = \frac{2\pi c_0}{m^2} \frac{1}{\omega^2 + \gamma^2} \delta(\omega - \omega') \tag{13.10}$$

The coefficient of the δ-function gives the power spectrum

$$S(\omega) = \frac{c_0}{m^2} \frac{1}{\omega^2 + \gamma^2} \tag{13.11}$$

13.2   Energy Balance

Multiply both sides of the Langevin equation [Equation (13.1)] with v, and take ensemble averages:

$$\frac{m}{2} \frac{d\langle v^2 \rangle}{dt} + m\gamma \langle v^2 \rangle = \langle vF \rangle \tag{13.12}$$

The average kinetic energy is defined by

$$K = \frac{m}{2} \frac{d\langle v^2 \rangle}{dt} \tag{13.13}$$

and evolves in time according to

$$\frac{dK}{dt} = \langle vF \rangle - 2\gamma K \tag{13.14}$$

The terms on the right side can be interpreted as follows:

$$\langle vF \rangle = \text{ average rate of work done on the particle}$$

$$2\gamma K = \text{ average rate of energy dissipation} \tag{13.15}$$

In steady state, when $dK/dt = 0$, these rates become equal to each other.
   To solve Equation (13.14), put

$$v = e^{-\gamma t} u \tag{13.16}$$

Substituting this into Langevin equation [Equation (13.1)] gives

$$m \frac{du}{dt} = e^{\gamma t} F \tag{13.17}$$

With the initial condition $v(0) = 0$, we have

$$u(t) = \frac{1}{m} \int_0^t dt' e^{\gamma t'} F(t')$$

$$v(t) = \frac{1}{m} \int_0^t dt' e^{\gamma (t'' - t)} F(t') \tag{13.18}$$

Multiplying the above by $F(t)$, and taking the ensemble average, we obtain

$$\langle vF \rangle = \frac{1}{m} \int_0^t dt' e^{\gamma(t'-t)} \langle F(t')F(t) \rangle$$

$$= \frac{1}{m} \int_0^t dt' e^{\gamma(t'-t)} c_0 \delta(t - t') = \frac{c_0}{2m} \tag{13.19}$$

The integral is ambiguous at face value, because $\delta(t - t')$ is nonvanishing only at the boundary $t' = t$. To render it well-defined, replace the δ-function by a Gaussian centered at $t' = t$. It is then clear that the integral covers half the area under the Gaussian, hence the factor 1/2. (See also Problem 13.2.)

Now Equation (13.14) becomes

$$\frac{dK}{dt} = \frac{c_0}{2m} - 2\gamma K \tag{13.20}$$

and the solution is

$$K(t) = \frac{c_0}{4m\gamma}(1 - e^{-2\gamma t}) \tag{13.21}$$

The energy balance is schematically illustrated in Figure 13.1. The mean-square velocity is given by

$$\langle v^2(t) \rangle = \frac{2K(t)}{m} = \frac{c_0}{2m^2\gamma}(1 - e^{-2\gamma t}) \tag{13.22}$$

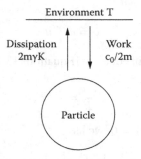

Figure 13.1 The Brownian particle comes to thermal equilibrium with the environment through energy exchange with the indicated rates. Here c_0 and γ are respectively the fluctuation and dissipation parameters in the Langevin equation, m is the mass, and K is the average kinetic energy.

13.3 Fluctuation-Dissipation Theorem

As $t \to \infty$, we have

$$K(\infty) = \frac{c_0}{4m\gamma} \qquad (13.23)$$

Equating $K(\infty) = \frac{1}{2}k_B T$ by the equipartition of energy, we obtain the *fluctuation-dissipation theorem*

$$c_0 = 2m\gamma k_B T \qquad (13.24)$$

The parameters c_0 and γ are different aspects of the same medium. This tells us how they must be related, as deduced from the requirement that the Brownian particle eventually reach thermal equilibrium with the medium.

A more suggestive form of the theorem can be obtained as follows. The Wiener–Kintchine theorem [Equation (12.15)] relates the power spectrum to the velocity correlation function:

$$S(\omega) = \int_0^\infty dt \, \langle v(t)v(0) \rangle \cos(\omega t) \qquad (13.25)$$

Using $S(\omega)$ from Equation (13.11), we obtain, after setting $\omega = 0$ and performing some substitutions,

$$\int_0^\infty dt \, \langle v(t)v(0) \rangle = k_B T \eta \qquad (13.26)$$

The left side is a manifestation of fluctuations, while the right side refers to dissipation.

13.4 Diffusion Coefficient and Einstein's Relation

Putting $v = dx/dt$, we rewrite the Langevin equation in the form

$$m\frac{d^2x}{dt^2} + m\gamma\frac{dx}{dt} = F \qquad (13.27)$$

The displacement $x(t)$ is expected to have a Gaussian distribution, since it is a stochastic variable. The variance of the distribution can be found as follows. Multiplying both sides of the equation by x, and taking ensemble averages, we have

$$\left\langle x\frac{d^2x}{dt^2} \right\rangle + \gamma \left\langle x\frac{dx}{dt} \right\rangle = 0 \qquad (13.28)$$

Here, we have put $\langle x F \rangle = 0$, by arguing that, while x changes sign under reflection, F on average does not. (See Problem 13.4 for actual calculation.) We can rewrite the above in the form

$$\frac{d^2}{dt^2}\langle x^2\rangle + \gamma \frac{d}{dt}\langle x^2\rangle - 2\langle v^2\rangle = 0 \tag{13.29}$$

From Equation (13.22), $\langle v^2 \rangle = \frac{c_0}{2m^2\gamma}(1 - e^{-2\gamma t})$. We neglect the exponentially decaying term, and take $\langle v^2 \rangle = \frac{c_0}{2m^2\gamma}$. Then the above equation yields

$$\langle x^2\rangle = \frac{c_0 t}{m^2\gamma^2} \tag{13.30}$$

This verifies that x obeys the law of diffusion, with variance increasing linearly with time. The diffusion coefficient D is obtained by equating the variance with $2Dt$:

$$D = \frac{c_0}{2m^2\gamma^2} \tag{13.31}$$

Using the fluctuation-dissipation theorem $c_0 = 2m\gamma k_B T$, we obtain Einstein's relation

$$D = \frac{k_B T}{m\gamma} \tag{13.32}$$

or, in terms of the mobility defined in Equation (13.3),

$$D = k_B T \eta \tag{13.33}$$

13.5   Transition Probability: Fokker–Planck Equation

The Fokker–Planck equation [Equation (12.46)] governs the time evolution of the probability distribution of v:

$$\frac{\partial P}{\partial t} + \frac{\partial}{\partial v}(AP) - \frac{1}{2}\frac{\partial^2}{\partial v^2}(BP) = 0 \tag{13.34}$$

where $P(v, t)dv$ is the probability of finding v within dv at time t. The functions $A(v)$ and $B(v)$ are related respectively to the mean and mean-square deviation of v over a small time interval Δt. We consider $t > 0$, with given initial condition at $t = 0$.

Integrating the Langevin equation [Equation (13.1)] over a small time interval Δt from a fixed initial value v, we have

$$\Delta v = -\gamma v \Delta t + \frac{1}{m}\int_t^{t+\Delta t} dt' \, F(t') \tag{13.35}$$

The second term cannot be equated with $F(t)\Delta t$, even though Δt is small, because $F(t)$ varies extremely rapidly. Using $\langle F \rangle = 0$, we obtain $\langle \Delta v \rangle = -\gamma v \Delta t$, hence

$$A(v) = -\gamma v \qquad (13.36)$$

The ensemble average of $(\Delta v)^2$ gives

$$\langle (\Delta v)^2 \rangle = \gamma^2 v^2 (\Delta t)^2 + \frac{1}{m^2} \int_t^{t+\Delta t} dt_1 \int_t^{t+\Delta t} dt_2 \, \langle F(t_1) F(t_2) \rangle \qquad (13.37)$$

The first term can be neglected, because it is of order $(\Delta t)^2$. Using Equation (13.4) we obtain

$$\langle (\Delta v)^2 \rangle = \frac{c_0}{m^2} \int_t^{t+\Delta t} dt_1 \int_t^{t+\Delta t} dt_2 \delta(t_1 - t_2) = \frac{c_0}{m^2} \Delta t \qquad (13.38)$$

Thus

$$B(v) = \frac{c_0}{m^2} \qquad (13.39)$$

If $v(t)$ were driven by a smoothly varying force, we would expect $\langle (\Delta v)^2 \rangle \sim (\Delta t)^2$; but the extremely short correlation time in the random force makes $\langle (\Delta v)^2 \rangle \sim \Delta t$.

The Fokker–Planck equation now reads

$$\frac{\partial P}{\partial t} - \gamma \frac{\partial}{\partial v}(vP) - \frac{c_0}{2m^2} \frac{\partial^2 P}{\partial v^2} = 0 \qquad (13.40)$$

The solution is

$$P(v, t) = \frac{1}{\sqrt{2\pi b}} \exp\left(-\frac{(v-a)^2}{2b}\right)$$

$$a = v_0 e^{-\gamma t}$$

$$b = \frac{c_0}{2m^2\gamma}(1 - e^{-2\gamma t}) \qquad (13.41)$$

where v_0 is an arbitrary constant. This is a Gaussian distribution for all t, as one expects from the central limit theorem. The time-dependent variance b agrees with Equation (13.22) from energy considerations.

The velocity distribution at $t = 0$ was $\delta(v - v_0)$. As time goes on, it becomes a Gaussian, whose width continues to broaden while the center shifts from v_0 toward zero. Memory of the initial state fades with a relaxation time γ^{-1}, and the distribution becomes Maxwell–Boltzmann.

13.6 Heating by Stirring: Forced Oscillator in Medium

We can heat a liquid by stirring it. A simple model for this is the Brownian motion of a forced harmonic oscillator in a medium. As illustrated schematically in Figure 13.2, energy input comes from both the driving force and the medium, and part of the

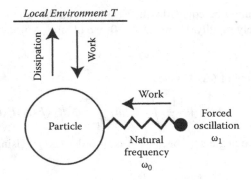

Figure 13.2 Brownian motion of a forced harmonic oscillator. The time-dependent driving heats the medium through "stirring," and modifies the fluctuation-dissipation theorem. The effect is most pronounced when the oscillator is driven at resonance.

work supplied by the driving force goes into heating the medium. This is indicated by a modification of the fluctuation-dissipation theorem.

The Langevin equation is

$$\frac{d^2x}{dt^2} + \gamma \frac{dx}{dt} + \omega_0^2 x = \frac{1}{m}(F + G) \tag{13.42}$$

where ω_0 is the natural frequency of the oscillator. The random force $F(t)$ is defined in Equation (13.4), and $G(t)$ is a driving force oscillating with frequency ω_1:

$$G(t) = b_0 \cos \omega_1 t \tag{13.43}$$

Let the Fourier transform of $x(t)$ be denoted by

$$x_\omega = \int_{-\infty}^{\infty} dt \, e^{i\omega t} x(t) \tag{13.44}$$

and that for $G(t)$ by

$$G_\omega = \pi b_0 \left[\delta(\omega - \omega_1) + \delta(\omega + \omega_1) \right] \tag{13.45}$$

The Fourier transform of the Langevin equation reads

$$\left(-\omega^2 - i\omega\gamma + \omega_0^2 \right) x_\omega = \frac{1}{m} (F_\omega + G_\omega) \tag{13.46}$$

and we immediately obtain

$$x_\omega = \frac{F_\omega + G_\omega}{m \left(\omega_0^2 - \omega^2 - i\omega\gamma \right)} \tag{13.47}$$

For the velocity $v(t) = dx/dt$, the Fourier transform is given by

$$v_\omega = -i\omega x_\omega \tag{13.48}$$

Going back to the Langevin equation (Equation [13.42]), we multiply both sides by dx/dt to get

$$m\frac{dx}{dt}\frac{d^2x}{dt^2} + m\omega_0^2\frac{dx}{dt}x + m\gamma\left(\frac{dx}{dt}\right)^2 = \frac{dx}{dt}(G+F)$$

$$\frac{d}{dt}\left(\frac{m}{2}v^2 + \frac{m\omega_0^2}{2}x^2\right) + m\gamma v^2 = v(G+F) \tag{13.49}$$

Taking ensemble averages, we have the equation for energy balance:

$$\frac{d}{dt}(K+U) = \langle vF\rangle + \langle vG\rangle - 2\gamma K \tag{13.50}$$

where K and U are respectively the average kinetic and potential energy:

$$K = \frac{m}{2}\langle v^2\rangle$$

$$U = \frac{m\omega_0^2}{2}\langle x^2\rangle \tag{13.51}$$

It is straightforward to calculate $\langle vF\rangle$ and $\langle vG\rangle$, though somewhat tedious. We merely quote the results here. Let an overhead bar denote the time average over periods of the forced oscillation:

$$\overline{\cos\omega_1 t} = \overline{\sin\omega_1 t} = 0$$

$$\overline{\cos^2\omega_1 t} = \overline{\sin^2\omega_1 t} = \frac{1}{2} \tag{13.52}$$

It can be shown that

$$\frac{d\overline{U}}{dt} = 0$$

$$\langle vF\rangle = \frac{c_0}{2m}$$

$$\overline{\langle vG\rangle} = \frac{\gamma b_0^2\omega_1^2}{2m}\frac{1}{\left(\omega_0^2-\omega_1^2\right)^2+\gamma^2\omega_1^2} \tag{13.53}$$

Thus

$$\frac{d\overline{K}}{dt} = \frac{c_0}{2m} + \frac{\gamma(b_0\omega_1)^2}{2m\left[\left(\omega_0^2-\omega_1^2\right)^2+\gamma^2\omega_1^2\right]} - 2\gamma\overline{K} \tag{13.54}$$

The average potential energy $\overline{U}$ does not participate in energy exchange with the environment. In equilibrium we should have $d\overline{K}/dt = 0$, and $\overline{K} = \frac{1}{2}k_BT$. Thus

$$c_0 + \frac{\gamma(b_0\omega_1)^2}{\left(\omega_0^2-\omega_1^2\right)^2+\gamma^2\omega_1^2} = 2m\gamma k_BT \tag{13.55}$$

which is the fluctuation-dissipation theorem. The second term arises from $\langle vG \rangle$, the work done by the driving force, and is maximum at the resonant frequency $\omega_1 = \omega_0$. In that case the fluctuation-dissipation theorem reduces to

$$c_0 + \frac{b_0^2}{\gamma} = 2m\gamma k_B T \quad \text{(at resonance: } \omega_1 = \omega_0\text{)} \tag{13.56}$$

This indicates that there is a net transfer of heat to the medium, due to the driving force.

By setting $\bar{K} = \frac{1}{2}k_B T$, we assume that the particle is in thermal equilibrium with the medium. In an infinite medium, overall equilibrium can never be reached. What happens is that local heating will create a temperature gradient in the vicinity of the particle, and the local temperature T is higher than that in the medium at infinity. The heated neighborhood will expand at a rate that depends on the thermal conductivity and specific heat of the medium.

Problems

13.1 The following integrals occur in calculating various correlation functions in Brownian motion using Fourier transforms. Verify the results using the techniques of complex contour integration.

$$\int_{-\infty}^{\infty} \frac{d\omega}{2\pi} \frac{1}{\omega^2 + \gamma^2} e^{i\omega t} = \frac{1}{2\gamma} e^{-\gamma |t|}$$

$$i \int \frac{d\omega}{2\pi} \frac{1}{\omega + i\gamma} e^{-i\omega t} = \begin{cases} e^{-\gamma t} & (t > 0) \\ 0 & (t < 0) \end{cases}$$

Note: The essential step in the complex integration is to decide whether to close the contour in the upper or lower half plane, so as to make the contribution of the infinite half circle negligible, and this depends on the sign of t.

13.2 Velocity-force correlation Calculate $\langle v(t)F(t')\rangle$. Take the limit $t' \to t$ and show $\langle vF \rangle = \frac{c_0}{2m}$.

Solution:

$$\langle u(\omega)f(\omega')\rangle = \frac{2\pi c_0 \delta(\omega + \omega')}{m(\gamma - i\omega)}$$

$$\langle v(t)F(t')\rangle = \int \frac{d\omega d\omega'}{(2\pi)^2} e^{-i\omega t} e^{-i\omega' t'} \frac{2\pi c_0 \delta(\omega + \omega')}{m(\gamma - i\omega)}$$

$$= \frac{i c_0}{m} \int \frac{d\omega}{2\pi} e^{-i\omega(t-t')} \frac{1}{\omega + i\gamma}$$

$$= \begin{cases} \frac{c_0}{m} e^{-\gamma(t-t')} & (t > t') \\ 0 & (t < t') \end{cases}$$

Taking the limit as $t \to t'$:

$$\langle v(t) F(t) \rangle = \frac{c_0}{2m} \qquad (13.57)$$

13.3 Displacement and causality

(a) Obtain $x(t)$ from its Fourier transform $x_\omega = \frac{i}{\omega} v_\omega$.

Solution:

$$x_\omega = \frac{i}{\omega} v_\omega = \frac{i}{m} \frac{F_\omega}{\omega(\gamma - i\omega)}$$

$$x(t) = \frac{i}{2\pi m} \int_{-\infty}^{\infty} \frac{d\omega}{\omega} \frac{F_\omega}{\gamma - i\omega} e^{-i\omega t}$$

This is ambiguous because of the pole at the origin. We will deal with this later.

(b) Calculate the correlation function $\langle x(t) F(t') \rangle$ from its Fourier transform $\langle x_\omega F_\omega \rangle$.

Solution:

$$\langle x_\omega F_\omega \rangle = \frac{i 2\pi c_0 \delta(\omega + \omega')}{m\omega(\gamma - i\omega)}$$

$$\langle x(t) F(t') \rangle = \frac{i c_0}{2\pi m} \int \frac{d\omega d\omega'}{\omega} \frac{\delta(\omega + \omega')}{\gamma - i\omega} e^{-i\omega t} e^{-i\omega' t'}$$

$$= -\frac{c_0}{2\pi m} \int \frac{d\omega}{\omega} \frac{1}{(\omega + i\gamma)} e^{-i\omega(t - t')}$$

The integral is ambiguous because of the pole at the origin. We must treat the pole in such a manner as to ensure causality:

$$\langle x(t) F(t') \rangle = 0 \quad (t < t')$$

That is, the particle cannot respond to a force applied in the future. This can be achieved by deforming the path above $\omega = 0$. Equivalently, we displace the pole to the lower half plane through the replacement

$$\frac{1}{\omega} \to \frac{1}{\omega + i\epsilon} \quad (\epsilon \to 0^+)$$

We then obtain, through contour integration,

$$\langle x(t) F(t') \rangle = \begin{cases} \frac{c_0}{\gamma}[1 - e^{-\gamma(t - t')}] & (t > t') \\ 0 & (t < t') \end{cases}$$

At equal time

$$\langle x(t) F(t) \rangle = 0$$

13.4 Variance of the displacement Calculate $\langle (\Delta x)^2 \rangle$ using $x(t)$ obtained in the last problem.

Solution:

$$x(t) = \frac{i}{2\pi m} \int_{-\infty}^{\infty} \frac{d\omega}{\omega + i\epsilon} \frac{F_\omega}{\gamma - i\omega} e^{-i\omega t} \quad (\epsilon \to 0^+)$$

Thus,

$$\langle (\Delta x)^2 \rangle \equiv \langle [x(t) - x(0)]^2 \rangle = \frac{2c_0}{\pi m^2} \int_{-\infty}^{\infty} d\omega \frac{\sin^2(\omega t/2)}{(\omega^2 + \epsilon^2)(\omega^2 + \gamma^2)}$$

where we can safely put $\epsilon = 0$. In the limit $t \to \infty$, we use the formula

$$\frac{\sin^2(wt)}{w^2} \xrightarrow[t \to \infty]{} \pi t \, \delta(w)$$

to get

$$\langle [x(t) - x(0)]^2 \rangle \xrightarrow[t \to \infty]{} \frac{c_0 t}{m^2 \gamma^2}$$

13.5 Show that the power spectrum of a harmonic oscillator in Brownian motion, with no driving force, is given by.

$$S(\omega) = \frac{c_0}{m^2} \frac{\omega^2}{\left(\omega^2 - \omega_0^2\right)^2 + (\omega\gamma)^2}$$

Solution:

$$x_\omega = -\frac{1}{m} \frac{F_\omega}{\omega^2 - \omega_0^2 + i\omega\gamma}$$

$$v_\omega = \frac{i}{m} \frac{\omega F_\omega}{\omega^2 - \omega_0^2 + i\omega\gamma}$$

$$\langle v_\omega v_{\omega'} \rangle = \frac{2\pi c_0}{m^2} \frac{\omega^2}{\left(\omega^2 - \omega_0^2\right)^2 + (\omega\gamma)^2} \delta(\omega + \omega')$$

The coefficient of the δ-function is $2\pi S(\omega)$.

Chapter 14

Quantum Statistics

14.1 Thermal Wavelength

Atoms in a gas are actually wave packets. They can be pictured as billiard balls at high temperatures, because their size is much smaller than the average interparticle distance. As the temperature decreases, however, the wave packets begin to spread, and when they begin to overlap with each other, specific quantum effects must be taken into account.

The spatial extension of a wave packet is governed by the deBroglie wavelength $\lambda_0 = h/p_0$, where p_0 is the average momentum. For a gas in equilibrium at temperature T, it is given through

$$\frac{p_0^2}{2m} = \frac{3}{2} k_B T \tag{14.1}$$

which gives $\lambda_0 = h/\sqrt{3mk_B T}$. We can make a wave packet by superimposing plane waves whose wavelengths lie in the neighborhood of λ_0. The spatial extension Δx and the momentum spread Δp must satisfy the uncertainly relation $\Delta x \Delta p \sim \hbar$. There is freedom to adjust Δx and Δp, but for reasonable choices λ_0 serves as an order-of-magnitude estimate of the degree of spatial localization.

We define the *thermal wavelength* by

$$\lambda = \sqrt{\frac{2\pi\hbar^2}{mk_B T}} \tag{14.2}$$

where the numerical factors are chosen to give a neater appearance to some formulas. For a system to be in the classical regime, λ should be much smaller than the average interparticle distance r_0. Since the latter is proportional to $n^{1/3}$, where n is the density, the condition can be stated as

$$n\lambda^3 \ll 1 \quad \text{(classical regime)} \tag{14.3}$$

Quantum effects become important at temperatures lower than the "degeneracy temperature" T_0 corresponding to

$$n\lambda^3 \approx 1 \quad \text{(onset of quantum effects)} \tag{14.4}$$

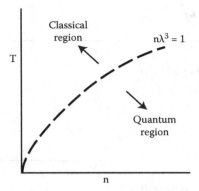

Figure 14.1 Classical and quantum regions in the temperature-density plane. The rough dividing line is $n\lambda^3 = 1$, where λ is the thermal wavelength.

At this point, the wave functions of different atoms begin to overlap, and we must treat the system according to quantum mechanics.

The condition $n\lambda^3 = 1$, or

$$n\left(\frac{2\pi\hbar^2}{mk_B T}\right)^{3/2} = 1 \tag{14.5}$$

defines a line in the T-n plane that serves as a rough division between the classical and quantum regimes, as indicated in Figure 14.1.

The degeneracy temperature T_0 is given by

$$k_B T_0 = \left(\frac{2\pi\hbar^2}{m}\right) n^{2/3} \tag{14.6}$$

Its value varies over a wide range for different physical systems, as indicated in Table 14.1, for example, at room temperature, a gas at STP can be described classically, whereas electrons in a metal are in the extreme quantum region. Liquid helium has a degeneracy temperature in between. At 2.17 K it makes a transition to a quantum phase that exhibits superfluidity.

TABLE 14.1 Quantum Degeneracy Temperatures

System	Density (cm³)	T_0 (K)
H₂ gas	2×10^{19}	5×10^{-2}
Liquid ⁴He	2×10^{22}	2
Electrons in metal	10^{22}	10^4

14.2 Identical Particles

In quantum mechanics atoms are identical, in the sense that the Hamiltonian is invariant under a permutation of their coordinates. This property has no analog in classical mechanics. In classical physics, particles have definite coordinates, and thus can be "tagged." In quantum mechanics, it is in principle impossible to tag them.

To illustrate the concept, let us consider two particles with respective coordinates $\mathbf{r}_1, \mathbf{r}_2$. The interchange of the coordinates can be represented by a permutation operation P on the wave function:

$$P\Psi(\mathbf{r}_1, \mathbf{r}_2) = \Psi(\mathbf{r}_2, \mathbf{r}_1) \tag{14.7}$$

Clearly $P^2 = 1$. The Hamiltonian is invariant under the permutation, namely, $PHP^{-1} = H$. This means that the operators P and H commute:

$$[P, H] = 0 \tag{14.8}$$

and we can simultaneously diagonalize these two operators. If Ψ is an eigenfunction of H with energy eigenvalue E, then so is $P\Psi$, with the same energy:

$$H\Psi = E\Psi$$
$$PH\Psi = EP\Psi$$
$$(PHP^{-1})(P\Psi) = E(P\Psi)$$
$$H(P\Psi) = E(P\Psi) \tag{14.9}$$

If the energy is not degenerated, then $P\Psi$ and Ψ must describe the same state, and $P\Psi$ can differ from Ψ at most by a normalization factor. Since $P^2 = 1$, we can choose that factor to be ± 1. Thus

$$\Psi(\mathbf{r}_1, \mathbf{r}_2) = \pm\Psi(\mathbf{r}_2, \mathbf{r}_1) \tag{14.10}$$

That is, the wave function must be either symmetric or antisymmetric under the interchange of particle coordinates. Particles with the symmetric property are said to obey *Bose statistics*, and are called *bosons*, while those with the antisymmetric property are said to obey *Fermi statistics*, and are called *fermions*. The wave function gives the probability amplitude of finding one particle at $\mathbf{r}_1$ and one particle at $\mathbf{r}_2$, but it cannot tell us which one.

The quantum-mechanical concept of identical particles affects the way we count states. Consider two identical particles localized at separated points A and B. Let $f(x)$ and $g(x)$ denote the one-particle wave function localized about A and B, respectively, with no overlap between the wave functions, as illustrated in Figure 14.2. If the particles were distinguishable, we would have two possible states $f(x_1)g(x_2)$ and $f(x_2)g(x_1)$, corresponding to particle 1 at A and particle 2 at B, and vice versa. For identical particles, however, there is only one state, with wave function

$$\Psi(x_1, x_2) = f(x_1)g(x_2) \pm f(x_2)g(x_1) \tag{14.11}$$

where the $+$ sign corresponds to Bose statistics, and the $-$ sign to Fermi statistics.

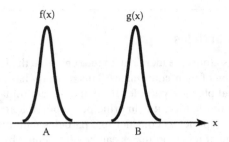

Figure 14.2 We can place two identical particles at the sites A and B, but it is impossible to specify which one is at A, and which at B.

The important point is that interchanging the coordinates of two particles does not change the state of the system. Thus, only half of the classical phase space of the two-particle system is relevant. For N particles, for which there are $N!$ permutations of the coordinates, only $1/N!$ of the phase space is relevant. This is the origin of "correct Boltzmann counting," which affects the entropy of the system.

14.3   Occupation Numbers

For N identical particles, the wave function $\Psi(\mathbf{r}_1, \ldots, \mathbf{r}_N)$ must be an even (odd) function under the interchange of any pair of coordinates for bosons (fermions). We can construct a complete set of N-body wave functions by first considering a complete orthonormal set of single-particle wave functions $u_\alpha(\mathbf{r})$, satisfying

$$\int d^3r \, u_\alpha^*(\mathbf{r}) u_\beta(\mathbf{r}) = \delta_{\alpha\beta} \tag{14.12}$$

where α is a single-particle quantum number. We then form a product of N of these wave functions, and symmetrize or antisymmetrize the product with respect to the coordinates $\{\mathbf{r}_1, \ldots, \mathbf{r}_N\}$. Equivalently, we can symmetrize (or antisymmetrize) with respect to the set of single-particle labels $\{\alpha_1, \ldots, \alpha_N\}$. The N-body wave function so obtained has the form

$$\Psi(\mathbf{r}_1, \ldots, \mathbf{r}_N) = \frac{1}{\sqrt{N!}} \sum_P \delta_P P[u_{\alpha_1}(\mathbf{r}_1) \cdots u_{\alpha_N}(\mathbf{r}_N)] \tag{14.13}$$

where P is a permutation of the set $\{\alpha_1, \ldots, \alpha_N\}$, and the sum extends over all $N!$ permutations. The signature factor δ_P is unity for bosons, while for fermions it is given by

$$\delta_P = \begin{cases} 1 & \text{if } P \text{ is an even permutation} \\ -1 & \text{if } P \text{ is an odd permutation} \end{cases} \quad \text{(for fermions)} \tag{14.14}$$

In the fermion case, the wave function is a determinant, called the *Slater determinant*:

$$\Psi(\mathbf{r}_1, \ldots, \mathbf{r}_N) = \frac{1}{\sqrt{N!}} \begin{vmatrix} u_1(\mathbf{r}_1) & u_1(\mathbf{r}_2) & \cdots & u_1(\mathbf{r}_N) \\ u_2(\mathbf{r}_1) & u_2(\mathbf{r}_2) & & u_2(\mathbf{r}_N) \\ \vdots & & & \\ u_N(\mathbf{r}_1) & u_N(\mathbf{r}_2) & & u_N(\mathbf{r}_N) \end{vmatrix} \qquad \text{(for fermions)}$$

$$(14.15)$$

We see immediately that Ψ vanishes whenever two single-particle wave functions are the same, for example, if $u_1(\mathbf{r}) \equiv u_2(\mathbf{r})$. This is the statement of the *Pauli exclusion principle*.

Because of indistinguishability, the N-body wave function is labeled by the set $\{\alpha_1, \ldots, \alpha_N\}$, in which the ordering of the set is irrelevant. To make this explicit, we specify the number of particles n_α in the state α. The number n_α is called the *occupation number* of the single-particle state α, with the allowed values

$$n_\alpha = \begin{cases} 0, 1, 2, \ldots, \infty & \text{(for bosons)} \\ 0, 1 & \text{(for fermions)} \end{cases}$$

$$(14.16)$$

For an N-particle system, they satisfy the condition

$$\sum_\alpha n_\alpha = N$$

$$(14.17)$$

A sum over states is a sum over all possible sets $\{n_\alpha\}$ satisfying Equation (14.17).

For free particles, it is convenient to choose the single-particle functions to be plane waves. The label α corresponds to the wave vector $\mathbf{k}$:

$$u_{\mathbf{k}}(\mathbf{r}) = \frac{1}{\sqrt{V}} e^{i\mathbf{k} \cdot \mathbf{r}}$$

$$(14.18)$$

In the thermodynamic limit, we can replace the sum over plane-wave states by an integral:

$$\sum_\alpha \rightarrow \int \frac{d^3 r \, d^3 p}{h^3}$$

$$(14.19)$$

This has the form of a classical phase-space integral, with unit specified by h^3. For plane-wave states, the state label α is specialized to momentum $\mathbf{p}$, and Equation (14.17) reads

$$\sum_{\mathbf{p}} n_{\mathbf{p}} = N$$

$$(14.20)$$

which in the thermodynamic limit becomes

$$\int \frac{d^3 r \, d^3 p}{h^3} n_{\mathbf{p}} = N$$

$$(14.21)$$

For a uniform gas, $n_{\mathbf{p}}$ is independent of the position, and the above becomes

$$\int \frac{d^3 p}{h^3} n_{\mathbf{p}} = \frac{N}{V} \tag{14.22}$$

14.4 Spin

For particles with intrinsic spin angular momentum, the single-particle states are labeled by wave vector $\mathbf{k}$ and spin state s:

$$\alpha = \{\mathbf{k}, s\} \tag{14.23}$$

The occupation number is denoted by $n_\alpha = n_{\mathbf{k}s}$. The total number of particles is given by

$$N = \sum_\alpha n_\alpha = \sum_{s=1}^{2S+1} \sum_{\mathbf{k}} n_{\mathbf{k}s} \tag{14.24}$$

For a free gas in equilibrium, $n_{\mathbf{k}s}$ is independent of the spin state, and the spin sum simply gives a factor $2S + 1$. Thus, the particles with different spin orientations form independent gases, each containing $N/(2S + 1)$ particles. This is why we can speak of "spinless fermions," from a mathematical point of view.

In nonrelativistic quantum mechanics, the wave function factors into a spatial part and a spin part:

$$u_{\mathbf{k}s}(\mathbf{r}) = \frac{1}{\sqrt{V}} e^{i\mathbf{k}\cdot\mathbf{r}} \chi_s(\sigma) \tag{14.25}$$

where σ is the spin coordinate. For the electron, which has spin $\frac{1}{2}$ (meaning that the spin angular momentum is $\hbar/2$,) the spin state has possible values $s = \pm 1$, corresponding to "up" or "down," and the spin coordinate is also two-valued: $\sigma = \pm 1$. The spin wave function is given by

$$\chi_s(\sigma) = \delta_{s\sigma} \tag{14.26}$$

which is often displayed in matrix form:

$$\chi_+ = \begin{pmatrix} 1 \\ 0 \end{pmatrix}, \quad \chi_- = \begin{pmatrix} 0 \\ 1 \end{pmatrix} \tag{14.27}$$

In the relativistic generalization, spin is mixed with spatial properties (Huang 1998). The wave function is no longer factorizable; but the enumeration of quantum numbers $\{\mathbf{k}, s\}$ remains the same, although extra quantum numbers appear referring to antiparticles.

14.5 Microcanonical Ensemble

The microcanonical ensemble is a collection of states with equal weight. For non-interacting particles, states specified by the occupations numbers $\{n_\alpha\}$ are equally probable, as long as they satisfy the constraints

$$\sum_\alpha n_\alpha = N$$

$$\sum_\alpha \epsilon_\alpha n_\alpha = E \qquad (14.28)$$

where ϵ_α is the energy of the single-particle state labelled by α. In the thermodynamic limit, the spectrum of states approach a continuum. As an aid to the counting of states, we divide the spectrum into discrete cells numbered $i = 1, 2, 3, \ldots$, with g_i states in cell i, as illustrated schematically in Figure 14.3. Each cell must contain a large number of states, but the cell should be small enough that the energies of the states are unresolved on a macroscopic scale. We denote the cell energy by ϵ_i, and refer to g_i as the cell degeneracy. The cell occupation numbers n_i satisfy the constraints

$$\sum_i n_i = N$$

$$\sum_i \epsilon_i n_i = E \qquad (14.29)$$

It should be noted that the division into cells is an imprecise coarse-graining process. The cell degeneracy g_i is an interim number that should disappear from the final expressions of physical quantities.

Because of coarse graining, the cell occupation $\{n_i\}$ corresponds to more than one quantum-mechanical state. The set $\{\bar{n}_i\}$ corresponding to the maximum number of quantum states is called the most probable set, and we assume that it describes thermal equilibrium. The development here bears a formal resemblance to the classical ensemble discussed in Section 5.8, but the fact that we have identical particles changes the rules for counting states; we have to use "quantum statistics."

To find the number of ways in which $\{n_i\}$ can be realized, we assign particles into single-particle quantum states, without paying attention to which particle goes into which cell, since the exchange of particles in different cells does not produce

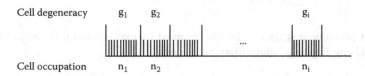

Figure 14.3 Grouping the near continuum of states into cells, in order to facilitate counting.

a new state. Thus, we can independently calculate the occupation of each cell, and multiply the results. The number of ways to obtain $\{n_i\}$ is therefore

$$\Omega\{n_i\} = \prod_j w_j(n_j) \qquad (14.30)$$

where $w_j(n_j)$ is the number of ways to put n_j particles into the jth cell, which contains g_j single-particle states.

The formula [Equation (14.30)] is very different from the corresponding classical formula [Equation (5.44)]. The classical way of counting in effect accepts all wave functions regardless of symmetry properties under the interchange of coordinates. The set of acceptable wave functions is far greater than the union of the two quantum cases. The classical way of counting is sometimes called "Boltzmann statistics," but it actually corresponds to no statistics.

14.6   Fermi Statistics

For fermions, each single-particle state can accommodate at most one particle, and thus each state is either occupied or empty. To place n_j fermions into g_j states, we pick the n_j occupied states out of a total of g_j states. The number of ways to achieve this is given by the binomial coefficient

$$w_j(n_j) = \frac{g_j!}{n_j!(g_j - n_j)!} \qquad (14.31)$$

Therefore

$$\Omega\{n_i\} = \prod_j \frac{g_j!}{n_j!(g_j - n_j)!} \qquad (14.32)$$

Assuming that n_j and g_j are large numbers, we use the Stirling approximation to obtain

$$\ln \Omega\{n_i\} = \sum_j [\ln g_j! - \ln n_j! - \ln(g_j - n_j)!]$$

$$\approx \sum_j [g_j \ln g_j - n_j \ln n_j - (g_j - n_j) \ln(g_j - n_j)] \qquad (14.33)$$

The most probable distribution $\{\bar{n}_i\}$ is obtained by maximizing this, subject to constraints. Using Lagrange multipliers, we require

$$\delta \left[\ln \Omega\{n_i\} + \alpha \sum_j n_j - \beta \sum_j \epsilon_j n_j \right] = 0$$

under independent variations of each n_j. This leads to

$$\sum_j \delta n_j [-\ln n_j + \ln(g_j - n_j) + \alpha - \beta \epsilon_j] = 0 \qquad (14.34)$$

Since the δn_j are independent and arbitrary, we must have

$$-\ln n_j + \ln(g_j - n_j) + \alpha - \beta \epsilon_j = 0 \qquad (14.35)$$

which gives

$$\bar{n}_j = \frac{g_j}{e^{-\alpha + \beta \epsilon_j} + 1} \qquad (14.36)$$

This is the Fermi distribution.

14.7   Bose Statistics

The jth cell is subdivided into g_j compartments by means of $g_j - 1$ partitions, and each compartment can hold any number of bosons. We can vary the particle numbers in the compartments by repositioning the partitions. The number of ways to populated the cell with n_j bosons is to throw the n_j particles into the cell, in any manner whatsoever, and then count the number of distinct configurations obtainable, when we permute the n_j particles together with the $g_j - 1$ partitions:

$$w_j(n_j) = \frac{(n_j + g_j - 1)!}{n_j!(g_j - 1)!} \qquad (14.37)$$

This gives

$$\Omega\{n_i\} = \prod_j \frac{(n_j + g_j - 1)!}{n_j!(g_j - 1)!}$$

$$\ln \Omega\{n_i\} \approx \sum_j [(n_j + g_j)\ln(n_j + g_j) - n_j \ln n_j - g_j \ln g_j] \qquad (14.38)$$

where we have put $g_j - 1 \approx g_j$. The most probable distribution is obtained through the condition

$$\sum_j \delta n_j [\ln(n_j + g_j - 1) - \ln n_j + \alpha - \beta \epsilon_j] = 0 \qquad (14.39)$$

with the result

$$\bar{n}_j = \frac{g_j}{e^{-\alpha + \beta \epsilon_j} - 1} \qquad (14.40)$$

This is the Bose distribution.

14.8   Determining the Parameters

The Fermi and Bose cases can be discussed together, by writing

$$n_j = \frac{g_j}{e^{-\alpha+\beta\epsilon_j} \pm 1} \quad \begin{array}{l} +: \text{Fermi} \\ -: \text{Bose} \end{array} \tag{14.41}$$

where we have omitted the bar over n_j for simplicity. The total number of particles is given by

$$N = \sum_j \frac{g_j}{e^{-\alpha+\beta\epsilon_j} \pm 1} \tag{14.42}$$

We may replace the sum over cells j by a sum over states $\mathbf{k}$, through the replacement

$$\sum_j g_j \rightarrow \sum_{\mathbf{k}} \tag{14.43}$$

Thus

$$N = \sum_{\mathbf{k}} \frac{1}{e^{-\alpha+\beta\epsilon_k} \pm 1} \tag{14.44}$$

where $\epsilon_k = \hbar^2 k^2/2m$, with $k = |\mathbf{k}|$. The cell degeneracy, which was a mathematical device, has now disappeared from the formula.

In the thermodynamic limit, we have

$$n = \int \frac{d^3k}{(2\pi)^3} \frac{1}{e^{-\alpha+\beta\epsilon_k} \pm 1} = \frac{8}{\pi^2} \int_0^\infty dk \frac{k^2}{e^{-\alpha+\beta\epsilon_k} \pm 1} \tag{14.45}$$

where n is the density. Let us make the substitution $\beta\epsilon_k = x^2$, or $k = x\sqrt{2m/\beta\hbar^2}$. Then

$$n = \frac{8}{\pi^2} \left(\frac{2m}{\beta\hbar^2} \right)^{3/2} \int_0^\infty dx \frac{x^2}{e^{-\alpha+x^2} \pm 1} \tag{14.46}$$

The density should remain finite in the limit $\beta \rightarrow 0$. This requires that the integral vanish, which in turn requires

$$e^{-\alpha} \xrightarrow[\beta\to 0]{} \infty \tag{14.47}$$

Thus the term ± 1 can be neglected in this limit, and we have

$$n_{\mathbf{k}} \xrightarrow[\beta\to 0]{} e^{\alpha-\beta\epsilon_k} \tag{14.48}$$

which corresponds to the Maxwell–Boltzmann distribution. Therefore we define the temperature T through

$$\beta = \frac{1}{k_B T} \tag{14.49}$$

Both the Bose and Fermi distributions approach the classical distribution in the high-temperature limit, because at high temperatures the particles tend to populate the excited states sparsely, and when there are few particles in a state, it matters little whether they are bosons or fermions.

The Lagrange multiplier α is determined by the condition (Equation [14.42]), which can be rewritten as

$$n\lambda^3 = \frac{4}{\sqrt{\pi}} \int_0^\infty dx \frac{x^2}{z^{-1}e^{x^2} \pm 1} \tag{14.50}$$

where $z = e^\alpha$ is called the *fugacity*. The *chemical potential* μ is defined through

$$z = e^{\beta\mu} \tag{14.51}$$

The occupation number for state $\mathbf{k}$ can now be written in the form

$$n_\mathbf{k} = \frac{1}{z^{-1}e^{\beta\epsilon_k} \pm 1} \tag{14.52}$$

This is the most probable occupation number, which is in principle different from the ensemble average. As in the classical case, the two coincide in the thermodynamic limit, but we postpone the proof to Chapter 15, where formal tools will be introduced.

14.9 Pressure

The pressure of the quantum ideal gas can be calculated using Equation (6.11), which is based on the particle flux impinging on a wall arising from a momentum distribution $f(\mathbf{p})$:

$$P = \int_{v_x > 0} d^3p (2mv_x)v_x \, f(\mathbf{p}) = m \int d^3p \, v_x^2 \, f(\mathbf{p}) \tag{14.53}$$

Using $f(\mathbf{p}) = h^{-3}n_\mathbf{p}$, and the fact that v_x^2 can be replaced by $\frac{1}{3}v^2$, we can rewrite

$$P = \frac{2}{3} \int \frac{d^3k}{(2\pi)^3} \frac{\epsilon_k}{z^{-1}e^{\beta\epsilon_k} \pm 1} \tag{14.54}$$

Changing the variable of integration such that $\beta\epsilon_k = x^2$, that is,

$$k = \sqrt{\frac{2mk_BT}{\hbar^2}} x = \frac{\sqrt{4\pi}x}{\lambda} \tag{14.55}$$

where λ is the thermal wavelength, we obtain

$$\frac{P}{k_BT} = \frac{1}{\lambda^3} \frac{8}{3\sqrt{\pi}} \int_0^\infty dx \frac{x^4}{z^{-1}e^{x^2} \pm 1} \tag{14.56}$$

The internal energy is given by

$$U = \sum_k \frac{\epsilon_k}{z^{-1} e^{\beta \epsilon_k} \pm 1} = V \int \frac{d^3 k}{(2\pi)^3} \frac{\epsilon_k}{z^{-1} e^{\beta \epsilon_k} \pm 1} \tag{14.57}$$

Comparison with Equation (14.54) yields the relation

$$PV = \frac{2}{3} U \tag{14.58}$$

This holds for the ideal Fermi and Bose gas, as well as the classical ideal gas. It depends only on the fact that the particles move in 3D with energy-momentum relation $\epsilon \propto (p_x^2 + p_y^2 + p_z^2)$.

14.10 Entropy

The entropy can be obtained from $S = k_B \ln \Omega$. We use Equation (14.33) for the Fermi case, and Equation (14.38) for the Bose case, to obtain

$$\frac{S}{k_B} = \sum_k \left[\frac{1}{\xi \pm 1} \ln(\xi \pm 1) \pm \frac{\xi}{\xi \pm 1} \ln \frac{\xi \pm 1}{\xi} \right] \tag{14.59}$$

where $\xi \equiv z^{-1} e^{-\beta \epsilon}$. Using the definition [Equation (14.57)] for U, and [Equation (14.42)] for N, we obtain

$$\frac{S}{k_B} = -N \ln z + \frac{U}{k_B T} \pm \sum_k \ln(1 \pm z e^{-\beta \epsilon}) \tag{14.60}$$

with $+$ for Fermi, and $-$ for Bose.

In the thermodynamic limit the last term becomes an integral, and we can make a partial integration. We show this explicitly for the Fermi case:

$$\frac{V}{2\pi^2} \int_0^\infty dk k^2 \ln \left(1 + z e^{-\beta \epsilon} \right)$$

$$= \frac{V}{2\pi^2} \left[\frac{k^3}{3} \ln(1 + z e^{-\beta \epsilon}) |_0^\infty - \int_0^\infty dk \frac{k^3}{3} \frac{\partial}{\partial k} \ln(1 + z e^{-\beta \epsilon}) \right]$$

$$= \frac{V}{6\pi^2} \frac{\beta \hbar^2}{m} \int_0^\infty dk \frac{k^4}{z^{-1} e^{\beta \epsilon} + 1} = \frac{2}{3} \frac{U}{k_B T} = \frac{PV}{k_B T} \tag{14.61}$$

The same final result holds for the Bose case. Therefore we have the relation

$$S = \frac{1}{T}(U + PV - N\mu) \tag{14.62}$$

where μ is the chemical potential defined in Equation (14.5). This verifies the thermodynamic relation $T^{-1} = \partial S / \partial U$.

14.11   Free Energy

We can identify the free energy as

$$A = N\mu - PV \tag{14.63}$$

for this obeys the Maxwell relations

$$\left(\frac{\partial A}{\partial V}\right)_{V,N} = -P$$

$$\left(\frac{\partial A}{\partial V}\right)_{V,T} = -\mu \tag{14.64}$$

Thus Equation (14.62) reduces to the thermodynamic relation

$$A = U - TS \tag{14.65}$$

From Equation (14.60) we obtain explicitly

$$A = k_B T \ln z \mp k_B T \sum_{\mathbf{k}} \ln(1 \pm z e^{-\beta\epsilon}) \tag{14.66}$$

where $+$ holds for the Fermi case, and $-$ for the Bose case.

14.12   Equation of State

The equation of state is no longer the simple relation $P = n k_B T$. It can be presented in parametric form by combining Equations (14.56) and (14.50):

$$\frac{\lambda^3 P}{k_B T} = \frac{8}{3\sqrt{\pi}} \int_0^\infty dx \frac{x^4}{z^{-1} e^{x^2} \pm 1}$$

$$n\lambda^3 = \frac{4}{\sqrt{\pi}} \int_0^\infty dx \frac{x^2}{z^{-1} e^{x^2} \pm 1} \tag{14.67}$$

The integrals on the right sides can be expanded in powers of z:

$$\frac{8}{3\sqrt{\pi}} \int_0^\infty dx \frac{x^4}{z^{-1} e^{x^2} \pm 1} = z \mp \frac{z^2}{2^{5/2}} + \frac{z^3}{3^{5/2}} \mp \cdots$$

$$\frac{4}{\sqrt{\pi}} \int_0^\infty dx \frac{x^2}{z^{-1} e^{x^2} \pm 1} = z \mp \frac{z^2}{2^{3/2}} + \frac{z^3}{3^{3/2}} \mp \cdots \tag{14.68}$$

It is convenient to introduce a class of Fermi functions $f_k(z)$ and Bose functions $g_k(z)$:

$$f_k(z) \equiv \sum_{\ell=1}^{\infty} (-1)^{\ell+1} \frac{z^\ell}{\ell^k}$$

$$g_k(z) \equiv \sum_{\ell=1}^{\infty} \frac{z^\ell}{\ell^k} \tag{14.69}$$

They obey the recursion relations

$$z\frac{d}{dz} f_k(z) = -f_{k-1}(z)$$

$$z\frac{d}{dz} g_k(z) = g_{k-1}(z) \tag{14.70}$$

with integral representations

$$\left.\begin{matrix} f_k(z) \\ g_k(z) \end{matrix}\right\} = \frac{2^{2k-1}}{\sqrt{\pi}} \frac{\Gamma\left(k-\frac{1}{2}\right)}{\Gamma(2k-1)} \int_0^\infty dx \frac{x^{2k-1}}{z^{-1}e^{x^2} \pm 1} \tag{14.71}$$

where $\Gamma(n)$ is the gamma function. We can then write the parametric equations of state for the Fermi case as

$$\frac{\lambda^3 P}{k_B T} = f_{5/2}(z)$$

$$\lambda^3 n = f_{3/2}(z) \tag{14.72}$$

and the Bose case as

$$\frac{\lambda^3 P}{k_B T} = g_{5/2}(z)$$

$$\lambda^3 n = g_{3/2}(z) \tag{14.73}$$

The Fermi functions have a singularity at $z = -1$, which lies outside of the physical region. The Bose functions have a singularity at $z = 1$, which leads to Bose–Einstein condensation, as we shall discuss in a later chapter.

We have obtained P and n as functions of the fugacity z, at a given temperature. What we need in applications is P as a function of n, and this cannot be obtained in closed form, but we can obtain it as a power series in n, which rapidly converges at high temperatures.

14.13   Classical Limit

In the high-temperature limit $\lambda \to 0$ (and hence $z \to 0$,) both quantum gases approach the classical behavior $P = nk_B T$. As the temperature decreases, their behaviors begin to diverge from each other, and when we reach the quantum region $n\lambda^3 \approx 1$

they become dramatically different. In the following, we examine the classical neighborhood in more detail.

Our object is to invert the equation

$$n\lambda^3 = z \mp \frac{z^2}{2^{3/2}} + \frac{z^3}{3^{3/2}} \mp \cdots \qquad (14.74)$$

in order to obtain z as a function of $n\lambda^3$. To do this by iteration, we take as first approximation

$$z \approx n\lambda^3 \qquad (14.75)$$

With this, the occupation number approaches the Maxwell–Boltzmann distribution:

$$n_{\mathbf{p}} \approx h^3 f(\mathbf{p}) \qquad (14.76)$$

and, as noted before, we recover the classical equation of state.

For the next approximation, we rewrite Equation (14.74) to second order, in the form

$$z = n\lambda^3 \pm \frac{z^2}{2^{3/2}} \qquad (14.77)$$

Substituting the first approximation $z = n\lambda^3$ on the right side, we obtain

$$z = n\lambda^3 \left(1 \pm \frac{n\lambda^3}{2^{3/2}} + \cdots \right)$$

$$\mu = k_B T \left[\ln(n\lambda^3) \pm \frac{n\lambda^3}{2^{3/2}} + \cdots \right] \qquad (14.78)$$

The second term is the first quantum correction, which has opposite signs for Fermi and Bose statistics. To this approximation, the equations state are

$$\frac{\lambda^3 P}{k_B T} = z \mp \frac{z^2}{2^{5/2}} + \cdots$$

$$= \left(n\lambda^3 \pm \frac{(n\lambda^3)^2}{2^{3/2}} \right) \mp \frac{1}{2^{5/2}} \left(n\lambda^3 \pm \frac{(n\lambda^3)^2}{2^{3/2}} \right)^2 + \cdots \qquad (14.79)$$

which leads to

$$\frac{P}{nk_B T} = 1 \pm 2^{-5/2} n\lambda^3 + \cdots \qquad (14.80)$$

where the upper sign is for Fermi statistics, and the lower sign for Bose statistics.

Compared to the classical gas of the same density and temperature, the pressure is larger for Fermi statistics, and smaller for Bose statistics. This indicates an effective repulsion between identical fermions, and attraction between identical bosons, even though there is no interparticle potential. These effective interactions arise from the correlations imposed by the symmetry of the quantum-mechanical wave function.

Problems

14.1 The thermal wavelength $\lambda = \sqrt{2\pi\hbar^2/mk_BT}$ applies only to nonrelativistic particles. For an ultra-relativistic particle, or a photon, show that it would be replaced by

$$L = \frac{\hbar c}{k_B T}$$

where c is the velocity of light. What would be the degeneracy temperature T_0?

14.2 Consider the temperature-density $(T\text{-}n)$ diagram of a free electron gas. Indicate the regions in which the system should be treated (a) relativistically; (b) quantum-mechanically.

14.3 At high temperatures the heat capacity C_V of a nonrelativistic monatomic gas approaches $\frac{3}{2}Nk_B$. Find the first quantum correction for both Fermi and Bose statistics. Expression the correction in terms of T/T_0, where T_0 is the degeneracy temperature.

14.4 For a gas at very low density or very high temperature, such that $\lambda^3 n \to 0$, the occupation number approaches

$$n_\lambda = z e^{-\beta\epsilon_\lambda}$$

independent of statistics. Although this has classical form, the energy spectrum ϵ_λ is still quantum mechanical.

(a) Show $z = Q/N$, where

$$Q = \sum_\lambda e^{-\beta\epsilon_\lambda}$$

This is the partition function to be discussed in more detail in Section 15.2.

(b) Show that the internal energy per particle is given by

$$\frac{U}{N} = -\frac{\partial}{\partial\beta}\ln Q$$

(c) For a polyatomic molecule, the energy has contributions from translational motion, and internal degrees of freedom such as rotations and vibrations. $\epsilon = \epsilon_{\text{trans}} + \epsilon_{\text{rot}} + \epsilon_{\text{vib}}$ show that partition function factorizes, and the specific heat capacity decomposes into a sum of terms:

$$Q = Q_{\text{trans}} Q_{\text{rot}} Q_{\text{vib}}$$

$$c_V = c_{\text{trans}} + c_{\text{rot}} + c_{\text{vib}}$$

where the subscripts refer to contributions from translational, rotational, and vibrational modes.

14.5 The translational energy is labeled by a wave vector $\mathbf{k}$, with $\epsilon_{\mathbf{k}} = \hbar^2 \mathbf{k}^2 / 2m$. Show that

$$Q_{\text{trans}} = \frac{V}{\lambda^3}$$

$$\frac{c_{\text{trans}}}{k} = \frac{3}{2}$$

14.6 A simple model for the rotational energy is $\epsilon_{\ell m} = \hbar^2 \ell(\ell + 1)/2I$, where $\ell = 0, 1, 2, \ldots, m = -\ell, \ldots, \ell$, and I is the moment of inertia. Thus

$$Q_{\text{rot}} = \sum_{\ell=0}^{\infty} (2\ell + 1)\, e^{-\beta \hbar^2 \ell(\ell+1)/2I}$$

(a) For $k_B T \ll \hbar^2/2I$ keep only the first two terms in Q_{rot}. Show

$$\frac{c_{\text{rot}}}{k} \approx \left(\frac{\beta \hbar^2}{I}\right)^2 e^{-\beta \hbar^2/I}$$

(b) For $k_B T \gg \hbar^2/2I$ approximate the sum over ℓ by an integral. Show

$$\frac{c_{\text{rot}}}{k} \approx 1$$

(c) Make a qualitative sketch of U/N and c_r as functions of temperature. Does c_{rot} approach the asymptotic value from above or from below?

14.7 The vibrational energy is $\epsilon_n = \hbar\omega(n + \frac{1}{2})$, where $n = 0, 1, 2, \ldots$, and ω is the vibrational frequency.
  (a) Show

$$\frac{c_{\text{vib}}}{k_B} = \left(\frac{\beta \hbar \omega}{1 - e^{-\beta \hbar \omega}}\right)^2$$

Make a qualitative sketch of c_{vib} as a function of temperature.
  (b) Find the mean value $\langle n + \frac{1}{2} \rangle$ and mean-square fluctuation $\langle (n + \frac{1}{2})^2 \rangle - \langle n + \frac{1}{2} \rangle^2$.

14.8 On the basis of the specific heats calculated above, we can now understand why equipartition works only among degrees of freedoms excited, as illustrated in Figure 6.2. Reproduce that figure qualitatively, and relate the threshold temperatures to the parameters in the energy spectrum.

14.9 Consider anharmonic vibrations with energy $\epsilon_n = \hbar\omega[(n + \frac{1}{2}) + b(n + \frac{1}{2})^2]$, where $b \ll 1$. Find the vibrational specific heat to first order in b. Use the results from Problem 14.7(b).

Reference

Huang, K., *Quantum Field Theory: From Operators to Path Integrals*, Wiley, New York, 1998, Chapter 6.

Chapter 15

Quantum Ensembles

15.1 Incoherent Superposition of States

In the last chapter we introduced quantum statistics via the microcanonical ensemble for ideal gases. We now give a more general discussion.

A fundamental difference between classical and quantum mechanics is that, while classical mechanics uses real numbers, quantum mechanics uses complex numbers. This is so because quantum mechanics describes a system in terms of wave functions, which are complex numbers, with a modulus and a phase. The phase gives rise to the phenomenon of quantum interference, which has no classical counterpart.

A statistical ensemble in quantum mechanics is a collection of states of the system that are completely independent of one another. In this respect it is no different from a classical ensemble. But we must require that there be no quantum interference among members of the ensemble. What is the wave function of a system described by such an ensemble?

Consider a system interacting with an environment, which we regard as a heat reservoir. Let $\Psi(x, X; t)$ denote the instantaneous wave function of system plus environment, where x denotes the coordinates of the system, and X those of the environment. Suppose H is the Hamiltonian of the system (not including the environment), with eigenfunctions $\psi_n(x)$ and eigenvalues E_n:

$$H\psi_n = E_n\psi_n \tag{15.1}$$

We can expand Ψ as a linear superposition of these eigenstates:

$$\Psi(x, X; t) = \sum_n c_n(X, t)\psi_n(x) \tag{15.2}$$

with the normalization condition

$$\sum_n \int dX\, c_n^* c_n = 1 \tag{15.3}$$

The coefficients $c_n(X, t)$ contain all the dependences on the time t and the environmental variables X.

Quantum interference occurs because Ψ represents not a probability, but a probability amplitude. The probability density of finding x and X with their designated values is its squared modulus:

$$|\Psi(x, X; t)|^2 = \sum_{m,n} c_m^*(X, t) c_n(X, t) \psi_m^*(x) \psi_n(x) \tag{15.4}$$

Using the polar representation

$$c_n = r_n e^{i\phi_n} \tag{15.5}$$

in which the dependence on (X, t) is understood, we have

$$|\Psi(x, X; t)|^2 = \sum_{m,n} e^{i(\phi_n - \phi_m)} r_m^* r_n \psi_m^*(x) \psi_n(x) \tag{15.6}$$

The terms with $m \neq n$ represent the quantum interference.

If the coefficients c_n have definite relative phases, the wave function Ψ corresponds to a single pure quantum state. If, for some reason, the relative phases can be regarded as random numbers, then the interference vanishes on average. In that case, Ψ describes an incoherent superposition of pure states. Such an incoherent superposition defines a quantum ensemble. We argue that randomness may arise owing to the complexity of the environment.

For an environment with many degrees of freedom, the relative phase $\phi_n(X, t) - \phi_m(X, t)$ is an extremely sensitive function of X, and it also fluctuates very rapidly with time. Now, the relative phase has an effective range $[0, 2\pi]$. If it changes by an amount much greater than 2π, in response to small changes of X and t, then on average it becomes uniformly distributed over $[0, 2\pi]$. The average of $\exp(i(\phi_n - \phi_m))$ over small intervals in X and t will become zero. The net result is that $\Psi(x, X; t)$ becomes an *incoherent superposition* of states. It corresponds to a quantum ensemble of the states described by $\psi_n(x)$.

The argument given above also applies to an isolated macroscopic system, for we can focus on one part of the system, and view the rest as environment.

15.2   Density Matrix

In an incoherent superposition of states, the relative phases do not enter explicitly into any computations. It would be convenient to use a description that avoids mentioning the phases altogether, and the density matrix enables us to do that.

The quantum mechanical expectation value of any observable O is given by

$$(\Psi, O\Psi) = \sum_{n,m} c_n^* c_m O_{nm} \tag{15.7}$$

where O_{nm} is the matrix element:

$$O_{nm} = \int dX \psi_n^*(x) O \psi_m(x) \tag{15.8}$$

The thermal average of O is obtained by integrating the expectation value over the coordinates of the environment, and averaging the result over time:

$$\langle O \rangle = \overline{\int dX(\Psi, O\Psi)} = \sum_{n.m} O_{nm} \overline{\int dX\, c_n^* c_m} \tag{15.9}$$

where an overhead bar denotes averaging over a time interval long compared with the relaxation time, but short on a macroscopic scale. The *random phase assumption* says

$$\overline{\int dX\, c_n^*(X, t) c_m(X, t)} = 0 \quad \text{if } n \neq m \tag{15.10}$$

The thermal average now takes the form

$$\langle O \rangle = \frac{\sum_{n.} \rho_n O_{nn}}{\sum_{n.} \rho_n} \tag{15.11}$$

where

$$\rho_n = \overline{\int dX |cn|^2}$$

This is the relative weight of the state ψ_n in the ensemble, and is the quantum analog of the classical density function that defines the ensemble.

We define an operator ρ whose eigenvalues are ρ_n with respect to the ψ_n basis:

$$\langle n|\rho|m \rangle = \delta_{nm} \rho_n \tag{15.12}$$

This is called the *density operator*. In the ψ_n basis it is represented by the matrix

$$\rho = \begin{pmatrix} \rho_1 & 0 & 0 & \\ 0 & \rho_2 & 0 & \\ 0 & 0 & \rho_3 & \\ & & & \ddots \end{pmatrix} \tag{15.13}$$

which is called the *density matrix*. If we change the basis, the matrix will become nondiagonal, but of course the operator itself remains unchanged.

The thermal average can now be written as

$$\langle O \rangle = \frac{\text{Tr}(\rho O)}{\text{Tr}\,\rho} \tag{15.14}$$

where Tr denotes the sum of diagonal elements. This definition is independent of matrix representation, due to the elementary property

$$\text{Tr}(AB) = \text{Tr}(BA) \tag{15.15}$$

A change of basis replaces all matrices M by SMS^{-1}, where S is the transformation effecting the change. Thus

$$\text{Tr}(SMS^{-1}) = \text{Tr}(MS^{-1}S) = \text{Tr}(M). \tag{15.16}$$

Thus we can calculate $\text{Tr}(\rho O)$ as the sum of diagonal elements of ρO in any matrix representation, including one in which ρ is not diagonal.

15.3 Canonical Ensemble (Quantum-Mechanical)

The canonical ensemble corresponds to the choice

$$\rho_n = e^{-\beta E_n} \tag{15.17}$$

The density operator is given by

$$\rho = \sum_n e^{-\beta E_n} |n\rangle \langle n| = e^{-\beta H} \sum_n |n\rangle \langle n|$$

$$= e^{-\beta H} \tag{15.18}$$

where we have used the completeness condition $\sum_n |n\rangle \langle n| = 1$. The thermal average now takes the form

$$\langle O \rangle = \frac{\text{Tr}(O e^{-\beta H})}{\text{Tr} \, e^{-\beta H}} \tag{15.19}$$

The quantum partition function is defined as

$$Q_N(V, T) = \text{Tr} \, e^{-\beta H} \tag{15.20}$$

where the trace extends over all states whose wave functions satisfy specified boundary conditions, with the correct symmetry as dictated by the statistics of the particles. The number of particles N enters through the Hamiltonian operator H, the volume V enters through the boundary condition imposed on the wave functions of the system, and of course the temperature enters through $\beta = 1/k_B T$. The internal energy is given by

$$U = \langle H \rangle = -\frac{\partial}{\partial \beta} \ln Q \tag{15.21}$$

The free energy $A(V, T)$ can be obtained through the formula

$$e^{-\beta A} = \text{Tr} \, e^{-\beta H} \tag{15.22}$$

and we can obtain all the thermodynamic functions through the Maxwell relations.

The quantum partition function can be written more explicitly as

$$Q = \text{Tr} \, e^{-\beta H} = \sum_\alpha \langle \alpha | e^{-\beta H} | \alpha \rangle \tag{15.23}$$

where $|\alpha\rangle$ is a member of any complete orthonormal set of states, with the properties

$$\langle \alpha | \beta \rangle = \delta_{\alpha\beta}$$

$$\sum_\alpha |\alpha\rangle \langle \alpha| = 1 \tag{15.24}$$

If we choose $|\alpha\rangle$ to be the energy eigenstates $|n\rangle$, then since $e^{-\beta H}|n\rangle = e^{-\beta E_n}|n\rangle$, we have

$$Q = \sum_n e^{-\beta E_n} \tag{15.25}$$

From a formal point of view, the energy representation appears to be a natural choice. However, it is not useful except for the ideal gas, because in general we are not able to diagonalize H. It practice, we choose the set $|\alpha\rangle$ that is most convenient for making approximations.

15.4 Grand Canonical Ensemble (Quantum-Mechanical)

The grand canonical ensemble is obtained from the canonical ensemble in the same method as in the classical case. The grand partition function is still given by Equation (9.4):

$$\mathcal{Q}(z, V, T) = \sum_{N=0}^{\infty} z^N Q_N(V, T) \tag{15.26}$$

except that here $Q_N(V, T)$ is the quantum partition function [Equation (15.20)]. The relation to thermodynamics remains the same as the classical case discussed in Section 9.4. Actual calculations may be very different. As we now illustrate, the grand partition function for the ideal gases can be explicitly calculated, although we cannot calculate the partition function.

The energy eigenvalues of the spinless quantum ideal gases are given by

$$E\{n\} = \sum_k n_k \epsilon_k \tag{15.27}$$

where n_k is the occupation number of the single-particle state with wave vector $\mathbf{k}$, energy $\epsilon_k = \hbar^2 k^2 / 2m$. The statistics is specified through the range:

$$n_k = \begin{cases} 0, 1 & \text{(Fermi)} \\ 0, 1, 2, \ldots & \text{(Bose)} \end{cases} \tag{15.28}$$

$n_k = 0, 1$ for fermions, and $n_k = 0, 1, \ldots, \infty$ for bosons. The canonical partition function is

$$Q_N(V, T) = \sum_{\substack{\{n\} \\ \Sigma_k n_k = N}} e^{-\beta E\{n\}} \tag{15.29}$$

where the sum over all possible sets of occupation numbers $\{n\}$ is subject to the condition $\sum_k n_k = N$. This constraint makes it impossible to calculate this sum explicitly.

However, we can calculate the grand partition function, which is given by

$$\mathcal{Q}(z, V, T) = \sum_{N=0}^{\infty} \sum_{\substack{\{n\} \\ \Sigma_k n_k = N}} z^{\Sigma_q n_q} e^{-\beta \Sigma_k n_k \epsilon_k} \tag{15.30}$$

where we have replaced the power N in z^N by $\Sigma_q n_q$. The double summation now collapses to a single sum over $\{n\}$, without constraint:

$$\sum_{N=0}^{\infty} \sum_{\substack{\{n\} \\ \Sigma_k n_k = N}} = \sum_{\{n\}} \qquad (15.31)$$

To see this, note that on the left side we first sum over sets of occupation numbers with a given N, and then sum over all N. On the right side, we have exactly the same sum, except that we don't bother to regroup the terms. We can further rewrite

$$z^{\Sigma_k n_k} e^{-\beta \Sigma_k n_k \epsilon_k} = \prod_q z^{n_q} \prod_k (e^{-\beta \epsilon_k})^{n_k} = \prod_k (z e^{-\beta \epsilon_k})^{n_k} \qquad (15.32)$$

Thus

$$\mathcal{Q}(z, V, T) = \sum_{\{n_0, n_1, \cdots\}} z^{\Sigma_q n_q} e^{-\beta \Sigma_k n_k \epsilon_k}$$

$$= \sum_{n_0} (z e^{-\beta \epsilon_0})^{n_0} \sum_{n_1} (z e^{-\beta \epsilon_1})^{n_1} \cdots$$

$$= \begin{cases} \prod_k (1 + z e^{-\beta \epsilon_k}) & \text{(Fermi)} \\ \prod_k (1 - z e^{-\beta \epsilon_k})^{-1} & \text{(Bose)} \end{cases} \qquad (15.33)$$

The average occupation number is given by

$$\langle n_k \rangle = \frac{1}{\mathcal{Q}(z, V, T)} \sum_{\{n\}} n_k z^{\Sigma_k n_k} e^{-\beta \Sigma_k n_k \epsilon_k} = -\frac{1}{\beta} \frac{\partial}{\partial \epsilon_k} \ln \mathcal{Q}(z, V, T)$$

$$= \frac{1}{z^{-1} e^{-\beta \epsilon_k} \pm 1} \qquad \begin{matrix} (+ : \text{Fermi}) \\ (- : \text{Bose}) \end{matrix} \qquad (15.34)$$

The equation of state in parametric form is

$$\frac{P}{k_B T} = \frac{1}{V} \ln \mathcal{Q}(z, V, T) = \pm \frac{1}{2\pi^2} \int_0^{\infty} dk k^2 \ln(1 \pm z e^{-\beta \epsilon_k})$$

$$n = \frac{1}{V} \sum_k \langle n_k \rangle = \frac{1}{2\pi^2} \int_0^{\infty} dk k^2 \frac{1}{z^{-1} e^{-\beta \epsilon_k} \pm 1} \qquad (15.35)$$

where the upper $\pm$ sign corresponds to Fermi statistic, and the lower sign to Bose statistics. We have taken the infinite-volume limit $V \to \infty$.

Making a partial integration using

$$\int dk k^2 f(k) = \frac{k^3}{3} f(k) - \int dk \frac{k^3}{3} \frac{\partial}{\partial k} f(k) \qquad (15.36)$$

we obtain

$$P = \frac{2}{3} \int \frac{d^3k}{(2\pi)^3} \frac{\epsilon_k}{z^{-1} e^{\beta \epsilon_k} \pm 1} = \frac{2}{3} U \qquad (15.37)$$

The equation of state obtained here, in the grand canonical ensemble, agrees with that in Chapter 14 obtained in the micocanonical ensemble.

15.5 Occupation Number Fluctuations

To calculate the mean-square fluctuation in the average occupation number of the ideal gases, we start with the definition

$$\langle n_k \rangle = \frac{1}{Q} \sum_{\{n_k\}} n_k e^{-\beta \sum_k n_k (\epsilon_k - \mu)} \qquad (15.38)$$

where we have set $z = e^{\beta \mu}$. Differentiating both sides with respect to ϵ_k, we have

$$\frac{\partial}{\partial \epsilon_k} \langle n_k \rangle = -\frac{\beta}{Q} \sum_{\{n_k\}} n_k^2 e^{-\beta \sum_k n_k (\epsilon_k - \mu)} - \frac{1}{Q^2} \frac{\partial Q}{\partial \epsilon_k} \sum_{\{n_k\}} n_k e^{-\beta \sum_k n_k (\epsilon_k - \mu)} \qquad (15.39)$$

The last term on the right side is of the form

$$-\frac{\partial \ln Q}{\partial \epsilon_k} \langle n_k \rangle = \beta \langle n_k \rangle^2 \qquad (15.40)$$

Thus

$$\frac{\partial}{\partial \epsilon_k} \langle n_k \rangle = -\beta \langle n_k^2 \rangle + \beta \langle n_k \rangle^2 \qquad (15.41)$$

or

$$\langle n_k^2 \rangle - \langle n_k \rangle^2 = -\frac{1}{\beta} \frac{\partial}{\partial \epsilon_k} \langle n_k \rangle$$

$$= \frac{e^{-\beta(\epsilon_k - \mu)}}{[1 \pm e^{-\beta(\epsilon_k - \mu)}]^2} \quad \begin{matrix} (- : \text{Fermi}) \\ (+ : \text{Bose}) \end{matrix} \qquad (15.42)$$

This can also be rewritten as

$$\langle n_k^2 \rangle - \langle n_k \rangle^2 = \langle n_k \rangle \mp \langle n_k \rangle^2 \quad \begin{matrix} (- : \text{Fermi}) \\ (+ : \text{Bose}) \end{matrix} \qquad (15.43)$$

The result for Fermi statistics is obvious, since in this case $n_k^2 = n_k$. Thus, the left side is equal to $\langle n_k \rangle - \langle n_k \rangle^2$. On the other hand, the Bose case shows unusually strong fluctuations. These have been experimentally observed, as we discuss next.

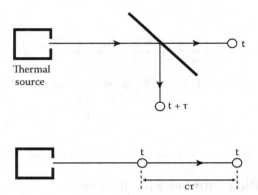

Figure 15.1 In the experiment of Hanbury Brown and Twiss, a light beam from a thermalized source is split, and the intensities are measured at the two ends with time difference τ. This is equivalent to taking instantaneous measurements along an unsplit beam at a distance $c\tau$ apart, where c is the velocity of light.

15.6 Photon Bunching

According to Equation (15.43), the fractional fluctuation in the occupation number in a gas of thermalized bosons is given by

$$\frac{\langle n_{\mathbf{k}}^2 \rangle - \langle n_{\mathbf{k}} \rangle^2}{\langle n_{\mathbf{k}} \rangle^2} = \frac{1}{\langle n_{\mathbf{k}} \rangle^2} + 1 \tag{15.44}$$

which is always greater than 100%. Such large fluctuations were indeed observed in a classic experiment by Hanbury Brown and Twiss (Hanbury Brown and Twiss 1957), which is regarded as the harbinger of quantum optics.

The experiment observes the self-interference in a beam of photons emerging from a thermal source, as schematically illustrated in Figure 15.1. A half-silvered mirror splits the photon beam in two. The intensity of one of the beams is measured at time t, while the other is measured at time $t + \tau$. The product of the intensities is then averaged over t. This is equivalent to simultaneously measuring the intensities in a single beam at two different points separated by distance $c\tau$, and averaging the product of the intensities over time.

We shall not go into quantitative details, but only report that photons were found to arrive in bunches. Since the intensity at a particular location gives the photon occupation number in the thermal source at an earlier time, the bunching of photons indicates that the occupation number strongly fluctuates in time.

Edward Purcell (1956) gives an intuitive explanation of the intensity fluctuations, which we paraphrase:

> Any real photon beam cannot be exactly monochromatic, for that would correspond to an infinite plane wave filling all space. A near-monochromatic beam consists of a train of pulses, each containing one

photon, with length of order $c/\Delta\nu$, where $\Delta\nu$ is the uncertainty in frequency. For small $\Delta\nu$ the pulses can be very long, and they come at random times. Occasionally two pulses will overlap, and when that happens one gets either 0 or 4 photons. This is why the photons appear bunched together.

The strong fluctuations can also be explained in the context of classical electromagnetic waves, and this raises the question whether it has anything to do with Bose statistics. The answer is yes, for it is Bose statistics that enables photons to exist in the same state, in sufficiently large numbers to produce a classical electromagnetic wave.

It should be mentioned that a laser beam does not fluctuate in intensity; it is a coherent superposition of states instead of an incoherent ensemble. To achieve coherence, the states in the superposition must contain varying numbers of photons.

Problems

15.1 To illustrate the difference between a pure state and an ensemble, consider an atom with two spin states "up" and "down":

$$\chi_1 = \begin{pmatrix} 1 \\ 0 \end{pmatrix}, \quad \chi_2 = \begin{pmatrix} 0 \\ 1 \end{pmatrix}$$

Consider the spin wave function obtained by superposing the two states:

$$\psi = \alpha\chi_1 + \beta\chi_2$$

where α and β are complex coefficients satisfying $|\alpha|^2 + |\beta|^2 = 1$. As long as α and β are definite complex numbers, this describes a state whose spin points along some direction rotated from the original axis of quantization. A beam of atoms having this spin wave function is 100% polarized along some direction. If, however, the relative phase between α and β is randomized, the beam becomes partially polarized, with a fraction $|\alpha|^2$ along the up axis, and a fraction $|\beta|^2$ along the down axis.

Let the scattering cross sections for a polarized beam be respectively σ_1 and σ_2, for 100% polarization along the up and down axes. Give the scattering cross section for a partially polarized beam in terms of a density matrix.

15.2 The one-dimensional harmonic oscillator has Hamiltonian

$$H = \frac{p^2}{2m} + \frac{1}{2}m\omega^2 q^2$$

where q is the coordinate, and p is the momentum.

(a) Calculate the classical partition function, taking the phase-space element to be $dpdq/\tau$, where τ is an arbitrary unit.

(b) In quantum mechanics, the eigenvalues of the Hamiltonian are

$$E_n = \hbar\omega \left(n + \frac{1}{2} \right)$$

where $n = 0, 1, 2, \ldots$ labels the quantum states. Show that the quantum partition function is

$$Q = \frac{e^{-\beta\hbar\omega/2}}{1 - e^{-\beta\hbar\omega}}$$

(c) Compare Q in the high-temperature limit ($\beta \to 0$) with the classical partition function, and show that the unit for the classical phase space is $\tau = h$, Planck's constant.

15.3 In quantum mechanics the kinetic energy K and the potential energy V do not commute with each other. This poses a challenge in calculating the partition function $\text{Tr}\, e^{-\beta(K+V)}$, because $e^{-\beta(K+V)} \neq e^{-\beta K} e^{-\beta V}$. According to the *Campbell-Baker-Hausdorff formula*, we have an infinite-product expansion

$$e^{-\beta(K+V)} = e^{-\beta K} e^{-\beta V} e^{-\beta^2 [K,V]/2} \cdots$$

where $[K, V] = KV - VK$, and the dots represent exponential factor involving higher powers of β and repeated commutators. Verify this formula to order β^2 by expanding both sides in powers of β.

15.4 Consider a system of N noninteracting spins, whose energies in a magnetic field B are given by $\pm \mu_0 B$. Ignore translational motion.
  (a) Calculate the partition function Q_N.
  (b) Calculate the average magnetic moment $\langle M \rangle$.
  (c) Find the mean-square fluctuation $\langle M^2 \rangle - \langle M \rangle^2$.

15.5 Consider an ideal Fermi gas of N spinless particles, with single-particle states labeled by a set of quantum numbers $\{\lambda\}$. Denote the energy levels by ϵ_λ.
  (a) Suppose the energy spectrum consists of a bound state of energy $-B$ that is g-fold degenerate, plus the usual free-particle energy spectrum $\epsilon_k = \hbar^2 k^2 / 2m$. Assume $g = aV$, where V is the volume, and a is a constant. Give the equation of state in parametric form.
  (b) Find z at high temperatures, as a function of temperature T and density n, to lowest order.
  (c) In the same approximation, find the pressure P, and the densities n_b and n_f of bound and free particles, respectively.

15.6 For an ideal quantum gas show that

$$\langle n_k n_p \rangle - \langle n_k \rangle \langle n_p \rangle = -\frac{1}{\beta} \frac{\partial}{\partial \epsilon_k} \langle n_p \rangle \quad (k \neq p)$$

Since $\langle n_p \rangle$ does not depend on ϵ_k, this gives

$$\langle n_k n_p \rangle = \langle n_k \rangle \langle n_p \rangle \quad (k \neq p)$$

Hint: Differentiate the grand partition function with respect to ϵ_p, and then ϵ_k.

15.7 We have obtained the fluctuation for the occupation of a single state. For a macroscopic system, the energy levels approach a continuum in the limit of infinite volume, and it is more relevant to consider a group of states. Let σ be the occupation number of a group G:

$$\sigma = \sum_{k \in G} n_k$$

Show the following relation, with help from the result of the last problem:

$$\langle \sigma^2 \rangle - \langle \sigma \rangle^2 = \langle \sigma \rangle \mp \sum_{k \in G} \langle n_k \rangle^2 \qquad \begin{array}{l} (\ - \ : \text{Fermi}) \\ (\ + \ : \text{Bose}) \end{array}$$

When no single state is macroscopically occupied, the left side is of order V^2, while the right side is of order V, and this represents a normal fluctuation. The exception occurs in Bose–Einstein condensation (See Chapter 18).

References

Hanbury Brown, R. and R.O. Twiss, *Nature*, **177**:27 (1957).

Purcell, E., *Nature*, **178**:1449 (1956).

Chapter 16

The Fermi Gas

16.1 Fermi Energy

In the low-temperature limit $T \to 0$, the Fermi distribution [Equation (14.36) or (15.34)] has the behavior

$$n_{\mathbf{k}} = \frac{1}{e^{\beta(\epsilon_k - \mu)} + 1} \xrightarrow[T \to 0]{} \begin{cases} 0 & \text{if } \epsilon_k > \mu \\ 1 & \text{if } \epsilon_k < \mu \end{cases} \qquad (16.1)$$

where $\beta = (k_B T)^{-1}$. Using the step function

$$\theta(x) = \begin{cases} 1 \text{ if } x > 0 \\ 0 \text{ if } x < 0 \end{cases} \qquad (16.2)$$

we write

$$n_{\mathbf{k}} \xrightarrow[T \to 0]{} \theta(\mu - \epsilon_k) \qquad (16.3)$$

This means that all states with energy below the *Fermi energy*

$$\epsilon_F = \mu(n, 0) \qquad (16.4)$$

are occupied, and all those above are empty. In momentum space the occupied states lie within the Fermi sphere of radius p_F, as illustrated in Figure 16.1. Such a condition is called "quantum degeneracy."

We can determine the Fermi energy through the condition

$$N = \sum_{\substack{\text{states with} \\ \epsilon < \epsilon_F}} 1 \qquad (16.5)$$

For free fermions of spin S the condition gives

$$N = \frac{V(2S + 1)}{(2\pi)^3} \frac{4\pi}{3} k_F^3 \qquad (16.6)$$

where $k_F = p_F / \hbar$ is the Fermi wave number. Thus

$$n = \frac{(2S + 1)k_F^3}{6\pi^2} \qquad (16.7)$$

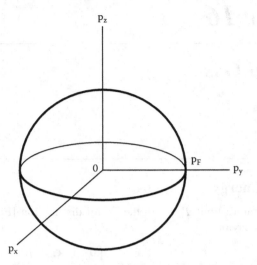

Figure 16.1 At absolute zero, all states within the Fermi sphere in momentum are occupied, and all others are empty. The radius of the sphere is the Fermi momentum $p_F = \hbar k_F$, where k_F is the Fermi wave vector. The Fermi energy is $\epsilon_F = p_F^2/2m$.

and the Fermi surface is described by

$$k_F = \left(\frac{6\pi^2 n}{2S+1} \right)^{1/3}$$

$$\epsilon_F = \frac{\hbar^2}{2m} \left(\frac{6\pi^2 n}{2S+1} \right)^{2/3} \tag{16.8}$$

16.2 Ground State

At absolute zero, when the system is in its quantum-mechanical ground state, the internal energy is given by

$$U_0 = \sum_{|\mathbf{k}|<k_F} \epsilon_k = \frac{V(2S+1)}{(2\pi)^3} \int_0^{k_F} dk(4\pi\, k^2) \frac{\hbar^2 k^2}{2m}$$

$$= \frac{V}{(2\pi)^3} \left(\frac{\hbar^2}{2m} \right) \frac{4\pi}{5} k_F^5 \tag{16.9}$$

On the other hand

$$N = \frac{V(2S+1)}{(2\pi)^3} \frac{4\pi}{3} k_F^3 \tag{16.10}$$

Thus the internal energy per particle at absolute zero is

$$\frac{U_0}{N} = \frac{3}{5}\frac{\hbar^2 k_F^2}{2m} = \frac{3}{5}\epsilon_F \tag{16.11}$$

This relation is independent of the spin of S.

We have shown in Section 14.9 that $PV = \frac{2}{3}U$ at any temperature. Thus, at absolute zero

$$P_0 = \frac{2}{5}n\epsilon_F \tag{16.12}$$

This zero-point pressure arises from the fact that there must be moving particles at absolute zero, since the zero-momentum state can hold only one particle of a given spin orientation. Taking a metal as a Fermi gas of electrons of spin 1/2 contained in a box, at density of $n \approx 10^{22}$ cm^{-3}, we find a zero-point pressure of the order of

$$P_0 \approx 10^{10} \text{ erg cm}^{-3} \approx 10^4 \text{ atm} \tag{16.13}$$

16.3 Fermi Temperature

At a finite temperature, the occupation number $n(\epsilon)$ as a function of the single-particle energy ϵ has the qualitative shape shown in the left panel in Figure 16.2. In the right panel, we show the momentum space, where particles within a layer beneath the Fermi surface are excited to a layer above, leaving "holes" beneath the Fermi surface.

The Fermi temperature T_F is defined by

$$\epsilon_F = k_B T_F \tag{16.14}$$

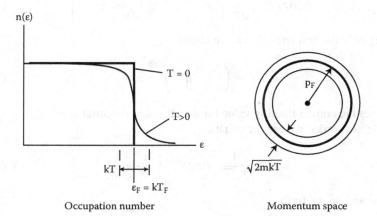

Occupation number   Momentum space

Figure 16.2 **Left:** Occupation number of a Fermi gas near absolute zero. **Right:** Particles are excited from beneath the Fermi surface to above, leaving holes behind.

For low temperatures $T \ll T_F$, the distribution deviates from that at $T = 0$ mainly in the neighborhood of $\epsilon = \epsilon_F$, in a layer of the thickness of order $k_B T$. That is, particles at energies of order $k_B T$ below the Fermi energy are excited to energies of order $k_B T$ above the Fermi energy. For $T \gg T_F$ the distribution is so broad that the Fermi surface disappears, and it approaches the classical Maxwell–Boltzmann distribution.

Electrons in a metal typically have $\epsilon_F \approx 2\,\mathrm{eV}$, which corresponds to $T_F \approx 2 \times 10^4$ K. Therefore, at room temperatures electrons are frozen below the Fermi level, except for a fraction $T/T_F \approx 0.015$. Since the average excitation energy per particle is $k_B T$, the internal energy is of order $(T/T_F)Nk_B T$, and the specific heat capacity tends to zero:

$$\frac{C}{Nk} \sim \frac{T}{T_F} \tag{16.15}$$

This is why the electronic contributions to the specific heat of a metal can be neglected at room temperature.

16.4   Low-Temperature Properties

The large z limit corresponds to $n\lambda^3 \gg 1$. At a fixed density, this corresponds to $T \ll T_F$, the degenerate limit. The condition for z is given in Section 14.12:

$$n\lambda^3 = f_{3/2}(z) \tag{16.16}$$

In the low-temperature region z is large, and we cannot use the power series for $f_{3/2}$. The relevant asymptotic formula is derived in the Appendix:

$$f_{3/2}(z) \approx \frac{4}{3\sqrt{\pi}} \left[(\ln z)^{3/2} + \frac{\pi^2}{8} \frac{1}{\sqrt{\ln z}} + \cdots \right] \tag{16.17}$$

Keeping only the first term above, we obtain

$$\ln z \approx \left(\frac{3\sqrt{\pi}}{4} n\lambda^3 \right)^{2/3} = \frac{T_F}{T} \tag{16.18}$$

This gives the correct limiting value for the chemical potential $\mu = \epsilon_F$.

To the next order, we use the equation

$$n\lambda^3 \approx \frac{4}{3\sqrt{\pi}} \left[(\ln z)^{3/2} + \frac{\pi^2}{8} \frac{1}{\sqrt{\ln z}} \right] \tag{16.19}$$

and rewrite it in the form

$$(\ln z)^{3/2} \approx \frac{3\sqrt{\pi}}{4} n\lambda^3 - \frac{\pi^2}{8} \frac{1}{\sqrt{\ln z}} \tag{16.20}$$

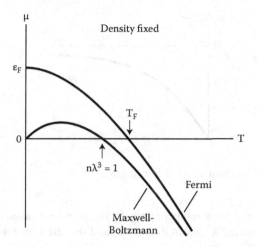

Figure 16.3 Chemical potential of the ideal Fermi gas. The Maxwell–Boltzmann case is shown for comparison.

Substituting the first approximation $\ln z = T_F / T$ on the right side, we then obtain

$$\ln z \approx \frac{T_F}{T}\left[1 - \frac{\pi^2}{12}\left(\frac{T}{T_F}\right)^2\right]$$

$$\mu \approx \epsilon_F\left[1 - \frac{\pi^2}{12}\left(\frac{T}{T_F}\right)^2\right] \tag{16.21}$$

Combining this with the high-temperature limit given in Section 14.13, we can sketch the qualitative behavior of the chemical potential as a function of temperature, as shown in Figure 16.3, where the Maxwell–Boltzmann curve refers to the high-temperature limit $\mu = k_B T \ln(n\lambda^3)$.

The internal energy is given by

$$U = \sum_{\mathbf{k}} \epsilon_{\mathbf{k}} n_{\mathbf{k}} = \frac{V}{(2\pi)^3} \frac{4\pi\hbar^2}{2m} \int_0^\infty dk\, k^4 n_{\mathbf{k}} \tag{16.22}$$

We make a partial integration:

$$
\begin{aligned}
U &= -\frac{V}{(2\pi)^3} \frac{4\pi\hbar^2}{2m} \int_0^\infty dk\, \frac{k^5}{5} \frac{\partial n_{\mathbf{k}}}{\partial k} \\
&= \frac{V}{(2\pi)^3} \frac{4\pi\hbar^2 \beta}{2m} \int_0^\infty dk\, \frac{k^5}{5} \frac{z^{-1} e^{\beta\epsilon}}{[z^{-1} e^{\beta\epsilon} + 1]^2} \frac{\partial \epsilon}{\partial k} \\
&= \frac{V\beta\hbar^4}{20\pi^2 m^2} \int_0^\infty dk\, \frac{k^6 e^{\beta(\epsilon-\mu)}}{[e^{\beta(\epsilon-\mu)} + 1]^2} \tag{16.23}
\end{aligned}
$$

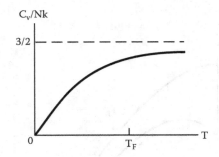

Figure 16.4 Specific heat of ideal Fermi gas.

where $\epsilon = \hbar^2 k^2/2m$. At low temperatures the integrand is peaked about $k = k_F$. We evaluate it by expanding k^6 about that point, and use the expansion for μ obtained earlier. The result is

$$U = \frac{3}{5}N\epsilon_F \left[1 + \frac{5\pi^2}{12}\left(\frac{T}{T_F}\right)^2 + \cdots \right] \tag{16.24}$$

from which we obtain the equation of state

$$P = \frac{2}{3}\frac{U}{V} = \frac{2}{5}n\epsilon_F \left[1 + \frac{5\pi^2}{12}\left(\frac{T}{T_F}\right)^2 + \cdots \right] \tag{16.25}$$

The heat capacity is given by

$$\frac{C_V}{Nk} = \frac{\pi^2}{2}\frac{T}{T_F} + \cdots \tag{16.26}$$

A qualitative plot of the specific heat is shown in Figure 16.4.

16.5 Particles and Holes

The absence of a fermion of energy ϵ, momentum $\mathbf{p}$, charge e, corresponds to the presence of a hole with

$$\text{Energy} = -\epsilon$$

$$\text{Momentum} = -\mathbf{p}$$

$$\text{Charge} = -e \tag{16.27}$$

Since the number of fermions in a quantum state is 0 or 1, the number of holes is 1 minus the number of fermions. It follows that, in a Fermi gas in thermal equilibrium,

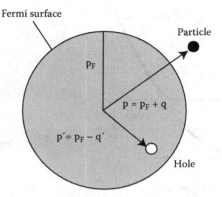

Figure 16.5 Particle excited to above the Fermi surface, leaving a hole beneath the Fermi surface.

the average occupation numbers for particle and holes are given by

$$\text{Particle:} \quad n_{\mathbf{p}} = \frac{1}{e^{\beta(\epsilon - \mu)} + 1}$$

$$\text{Hole:} \quad 1 - n_{\mathbf{p}} = \frac{1}{e^{\beta(\mu - \epsilon)} + 1} \tag{16.28}$$

where $\epsilon = \mathbf{p}^2/2m$. As illustrated in Figure 16.5, the number of excited particles is equal to the number of holes. When an excited fermion falls into an unoccupied state, it appears that a particle and a hole have annihilated each other. In this sense, a hole is an "antiparticle."

The concept of holes is useful only at low temperatures $T \ll T_F$, when there are relatively few of them below the Fermi surface, with the same number of excited particles above the Fermi surface. Only the holes and excited particles participate in thermal activities, while the rest of the fermions, which constitute the overwhelming majority, lie dormant beneath the Fermi surface. When the temperature is so high that there are a large number of holes, the Fermi surface is "washed out," and the system approaches a Maxwell–Boltzmann distribution of particles.

16.6 Electrons in Solids

An electron in the ionic lattice of a solid sees a periodic potential, and experiences partial transmission and reflection as it travels through the lattice. The reflected waves can interfere coherently with the original wave to produce complete destructive interference in certain energy ranges. Thus, energy gaps occur in the spectrum of the electron, and the lattice acts like a band-pass filter. A two-band spectrum is illustrated in Figure 16.6, with a "valence band" with maximum energy ϵ_0, separated by an

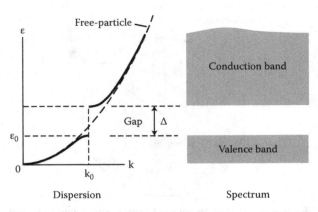

Figure 16.6 Band structure in the energy of an electron in a periodic potential.

energy gap Δ from the "conduction band" above it. Suppose the energy bands are filled with electrons. The two spin states of the electron only supply a factor of 2 in the counting of states. If the density is very low, we have a free Fermi gas with $\epsilon_F \ll \epsilon_0$, and the band structure has little relevance. Interesting phenomena happen when the valence is almost filled, or a bit overfilled.

Suppose at absolute zero the valence band is completely full, and the conduction band completely empty. Such a system is called a "natural semiconductor." As the temperature increases, thermal excitation will cause some electrons to be excited across the gap into the conduction band, leaving holes in the valence band. The particles and holes behave like free charge carriers, and the system exhibits electrical conductivity. As illustrated in Figure 16.6, the dispersion curve near the edges of the bands can be locally approximated by parabolas. An excited electron near the bottom conduction band has energy

$$\epsilon_c = \epsilon_0 + \Delta + \frac{\mathbf{p}^2}{2m_c} \tag{16.29}$$

where $\mathbf{p}$ is an effective momentum vector, and m_c is an effective mass. Similarly, the energy of an empty state just below the top of the valence band, has energy

$$\epsilon_v = \epsilon_0 - \frac{\mathbf{p}^2}{2m_v} \tag{16.30}$$

The densities of particles and holes are thus given by

$$n_{\text{part}} = 2 \int \frac{d^3p}{h^3} \frac{1}{e^{\beta(\epsilon_c - \mu)} + 1}$$

$$n_{\text{hole}} = 2 \int \frac{d^3p}{h^3} \frac{1}{e^{\beta(\mu - \epsilon_v)} + 1} \tag{16.31}$$

Assuming $\beta(\epsilon_c - \mu) \gg 1$, and $\beta(\mu - \epsilon_v) \gg 1$, we have

$$n_{\text{part}} \approx 2 \int \frac{d^3 p}{h^3} e^{-\beta(\epsilon_c - \mu)} = 2e^{-\beta(\epsilon_0 + \Delta - \mu)} \left(\frac{m_c k_B T}{2\pi \hbar^2} \right)^{3/2}$$

$$n_{\text{hole}} \approx 2 \int \frac{d^3 p}{h^3} e^{-\beta(\mu - \epsilon_v)} = 2e^{-\beta(\mu - \epsilon_0)} \left(\frac{m_v k_B T}{2\pi \hbar^2} \right)^{3/2} \qquad (16.32)$$

Overall electrical neutrality requires $n_{\text{part}} = n_{\text{hole}}$, which determines the chemical potential:

$$\mu = \epsilon_0 + \frac{\Delta}{2} + \frac{3}{4} k_B T \ln \frac{m_c}{m_v} \qquad (16.33)$$

The value at $T = 0$ gives the Fermi energy of $\epsilon_0 + \Delta/2$, which is located in the middle of the energy gap. The approximations made here are valid if $\beta \Delta \gg 1$. The particle and hole densities are now given by

$$n_{\text{part}} = n_{\text{hole}} = 2e^{-\beta \Delta/2} \left(\frac{\sqrt{m_c m_v} k_B T}{2\pi \hbar^2} \right)^{3/2} \qquad (16.34)$$

As an estimate, use the typical values $\Delta/k = 0.7\text{eV}$, and $m_c = m_v = m$, where m is the free electron mass. At room temperature $T = 300$ K, we find $n_{\text{part}} \approx 1.6 \times 10^{13}$ cm^{-3}, which is to be compared with a charge carrier density of 10^{20} for a metal. The electrical conductivity of a natural semiconductor is therefore negligible.

There are, however, gapless natural semiconductors with $\Delta = 0$, such as α-Sn and Hg-Te. In these cases the charge carrier densities become large, and depend on temperature like $T^{3/2}$.

16.7 Semiconductors

The electronic energy spectrum in a real solid rarely looks like that shown in Figure 16.6, because there always exist impurities in the lattice that can trap electrons into bound states. These impurities act as sources or sinks for electrons, and cases of practical interest are depicted in Figure 16.7.

In *n-type semiconductors* (n for negative), the bound levels lie below the bottom of the conduction band by an energy $\delta \ll \Delta$, and they are filled by electrons at absolute zero. As the temperature increases, these levels donate electrons to the originally empty conduction band. For this reason they are called *donor levels*.

In *p-type semiconductors* (p for positive), the bound levels lie above the top of the valence band by $\delta' \ll \Delta$, and are empty at absolute zero. As the temperature increases, they accept electrons excited from the valence band, thereby creating holes. For this reason they are called *acceptor levels*. We neglect excitations between bands.

For the n-type semiconductor, let us measure energy with respect to the bottom of the conduction band. Suppose there are n_D bound states per unit volume, of energy $-\delta$.

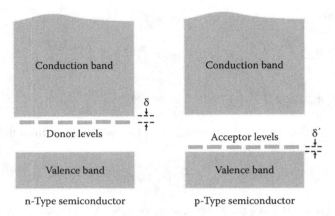

Figure 16.7 Impurities in the lattice of the solid can trap electrons into bound states, with energy levels in the band gap.

The density of electrons in the donor levels at temperature T is

$$n_{\text{donor}} = \frac{2n_D}{e^{-\beta(\delta+\mu)} + 1} \tag{16.35}$$

The density of electrons in the conduction band is given by n_{part} in Equation (16.32), with $\epsilon_0 + \Delta = 0$ in our convention:

$$n_{\text{part}} = \frac{2z}{\lambda^3}$$

$$\lambda = \sqrt{\frac{2\pi\hbar^2}{m_c k_B T}} \tag{16.36}$$

where $z = e^{\beta\mu}$ is the fugacity, and λ is the thermal wavelength with respect to the effective mass m_c. Since electrons in the conduction band were excited from the fully occupied donor levels, we must have $n_{\text{part}} + n_{\text{donor}} = 2n_D$, or

$$\frac{2z}{\lambda^3} + \frac{2n_D}{z^{-1}e^{-\beta\delta} + 1} = 2n_D \tag{16.37}$$

To solve for z, rewrite this in the form

$$e^{\beta\delta}z^2 + z - n_D\lambda^3 = 0 \tag{16.38}$$

The solution is

$$z = \frac{1}{2}e^{-\beta\delta}\left(\sqrt{4n_D\lambda^3 + 1} - 1\right) \tag{16.39}$$

which leads to

$$n_{\text{part}} = \frac{e^{-\beta\delta}}{\lambda^3}\left(\sqrt{4n_D\lambda^3 + 1} - 1\right) \tag{16.40}$$

As the temperature rises from absolute zero, the particles in the conduction band initially increase exponentially like $T^{3/4}e^{-\delta/k_B T}$, but flatten to a plateau at $2n_D$, when the donor levels are depleted. The particle density can be adjusted by changing n_D. In practice this is done through "doping"—mixing in varying amounts of impurity material.

P-type semiconductors can be discussed in the same manner, with holes replacing particles:

$$n_{\text{hole}} = \frac{e^{-\beta\delta'}}{\lambda^3}\left(\sqrt{4n_A\lambda^3 + 1} - 1\right) \tag{16.41}$$

where n_A is the number of acceptor states per unit volume, and $\lambda = \sqrt{2\pi\hbar^2/m_v k_B T}$.

Problems

16.1 A metal contains a high density of electrons, with interparticle distance of the order of 1 A. However, the mean-free-path of electrons at room temperature is very large, of the order of 10^4 A. This is because only a small fraction of electrons near the Fermi surface are excited. How does the mean-free-path depend on the temperature?

16.2 Model a heavy nucleus of mass number A as a free Fermi gas of an equal number of protons and neutrons, contained in a sphere of radius $R = r_0 A^{1/3}$, where $r_0 = 1.4 \times 10^{-13}$ cm. Calculate the Fermi energy and the average energy per nucleon in MeV.

16.3 Consider a relativistic gas of N particles of spin 1/2 obeying Fermi statistics, enclosed in volume V, at absolute zero. The energy-momentum relation is $E = \sqrt{(pc)^2 + (mc^2)^2}$, where m is the rest mass.

(a) Find the Fermi energy at density n.

(b) Define the internal energy U as the average of $E - mc^2$, and the pressure P as the average force per unit area exerted on a perfectly-reflecting wall of the container. Set up expressions for these quantities in the form of integrals, but you need not evaluate them.

(c) Show that $PV = 2U/3$ at low densities, and $PV = U/3$ at high densities. State the criteria for low and high densities.

(d) There may exist a gas of neutrinos (and/or antineutrinos) in the cosmos. (Neutrinos are massless fermions of spin 1/2.) Calculate the Fermi energy (in eV) of such a gas, assuming a density of one particle per cm^3.

16.4 Consider an ideal gas of nonrelativistic atoms at density n, with mass m, spin 1/2 and magnetic moment μ. The spin-up and spin-down atoms constitute two independent gases.

(a) In external magnetic field H at absolute zero, the atoms occupy all energy levels below a certain Fermi energy $\epsilon(H)$. Find the number $N_\pm$ of spin-up and spin-down atoms.

Hint: The energy of an atom is $\left(p^2/2m\right) \mp \mu H$.

(b) Find the minimum external field that will completely polarize the gas, as a function of the total density n.

16.5 Model a neutron star as an ideal Fermi gas of neutrons at absolute zero, in the gravitational field of a heavy center of mass M.

Show that the pressure P of a gas is in the gravitational field of a heavy mass M obeys the equation $dP/dr = -\gamma M\rho(r)/r^2$, where γ is the gravitational constant, r is the distance from the mass, and $\rho(r)$ is the mass density of the gas.

(b) Show that $P = a\rho^{5/3}$, where a is a constant, and find ρ as a function of distance from the center.

Hint: Consider how P and ρ depend on the chemical potential in a Fermi gas.

16.6 Consider an ideal Fermi gas of N electrons of mass m. In addition to the usual free-particle states, a single electron has N bound states that have the same energy $-\varepsilon$. (Each of these states can be occupied by one electron of given spin.)

(a) Write down expressions for N_b and N_f, the average number of bound and free particles in the gas.

(b) Write down the condition determining the fugacity $z = e^{\beta\mu}$.

(c) Find z as a function of temperature and density, assuming that $z \ll 1$. At what temperatures is this assumption valid?

(d) Find the density of free particles in the low-temperature and high-temperature limits.

Consider electrons in a metal to be an ideal Fermi gas of spin 1/2 particles.

16.7 At room temperature the distribution is close to that at absolute zero. There are relatively few electrons excited above the Fermi energy, leaving holes in the Fermi sea. Show that the probability $P(\Delta)$ of finding an electron with energy Δ above the Fermi energy is equal to the probability $Q(\Delta)$ of finding a hole at energy Δ below the Fermi energy.

16.8 Consider a two-dimensional Fermi gas consisting of N electrons of spin 1/2, confined in a box of area A.

(a) What is the density of single-particle states in momentum space?

(b) Calculate the density $D(\epsilon)$ of single particle states in energy, and sketch the result.

(c) Find the Fermi energy and Fermi momentum.

(d) Find the internal energy at $T = 0$ as a function of N and A.

(e) Find the surface tension S at $T = 0$ as a function of N and A.

Hint: $dE = TdS + SdA + \mu dN$.

(f) In 3D the chemical potential depends on T like T^2 at $T = 0$. Will the behavior in 2D stronger, weaker, or the same?

Chapter 17

The Bose Gas

17.1 Photons

Photons are the quanta of the electromagnetic field. They are bosons whose number is not conserved, for they may be created and absorbed singly. The Lagrange multiplier corresponding to total number is absent, and the chemical potential μ is zero. This means that the particles can disappear into the vacuum.

A photon has two spin states, corresponding to two possible polarizations, and travels at the speed of light c, with energy-momentum relation

$$\epsilon = cp \tag{17.1}$$

where p is the magnitude of the momentum $\mathbf{p}$. The wave number k and the frequency ω are defined through

$$p = \hbar k = \frac{\hbar \omega}{c} \tag{17.2}$$

Its intrinsic spin is $\hbar$, but has only two spin states corresponding to the states of left and right circular polarization. Any enclosure with walls made of matter must contain a gas of photons at the same temperature as the walls, due to emission and absorption of photons by the atoms in the walls. Such a system is called a "black-body cavity," and the photon it contains is called "black-body radiation."

The average occupation number of the state with momentum $\mathbf{p}$ is

$$n_{\mathbf{p}} = \frac{1}{e^{\beta cp} - 1} \tag{17.3}$$

The absence of the chemical potential makes thermodynamic calculations very simple. The average number of photons is

$$N = 2 \sum_{\mathbf{p}} n_{\mathbf{p}} = 2V \int \frac{d^3 p}{h^3} \frac{1}{e^{\beta cp} - 1} \tag{17.4}$$

where V is the volume of the system, and the factor 2 comes from the states of polarization. The density of photons is

$$n = \frac{2}{(2\pi)^3} \int_0^\infty dk \frac{4\pi k^2}{e^{\beta \hbar ck} - 1} = \frac{1}{\pi^2 c^3} \int_0^\infty d\omega \frac{\omega^2}{e^{\beta \hbar \omega} - 1} = \kappa \left(\frac{k_B T}{\hbar c} \right)^3 \tag{17.5}$$

with

$$\kappa = \frac{1}{\pi^2} \int_0^\infty dx \frac{x^2}{e^x - 1} = 4\zeta(3) \approx 0.23 \tag{17.6}$$

where $\zeta(z)$ is the Riemann zeta function.

The internal energy is given by

$$U = 2 \sum_{\mathbf{p}} \epsilon_{\mathbf{p}} n_{\mathbf{p}} = \frac{8\pi c V}{h^3} \int_0^\infty dp \frac{p^3}{e^{\beta c p} - 1}$$

$$= \frac{V\hbar}{\pi^2 c^3} \int_0^\infty d\omega \frac{\omega^3}{e^{\beta\hbar\omega} - 1} \tag{17.7}$$

which leads to *Stefan's law*

$$\frac{U}{V} = \sigma T^4$$

$$\sigma = \frac{\pi^2 k^4}{15(\hbar c)^3} \tag{17.8}$$

where σ is called *Stefan's constant*. The specific heat (per unit volume) is therefore

$$c_V = 4\sigma T^3 \tag{17.9}$$

The energy density can be expressed in the form

$$\frac{U}{V} = \int_0^\infty d\omega\, u(\omega, T) \tag{17.10}$$

where $u(\omega, T)$ is the energy density per unit frequency interval:

$$u(\omega, T) = \frac{\hbar}{\pi^2 c^3} \frac{\omega^3}{e^{\beta\hbar\omega} - 1} \tag{17.11}$$

This is the *Planck distribution*.

The pressure can be obtained through the method described in Section 6.2 modified for photon kinematics. Consider photons reflecting from a wall normal to the x axis. For photons of momentum $\mathbf{p}$, which makes an angle θ with the x axis, the flux of photons is $c \cos\theta$, and the momentum imparted to the wall per reflection is $2p \cos\theta$. Therefore the pressure is given by

$$P = 2 \int_{p_x > 0} \frac{d^3 p}{h^3} 2pc n_{\mathbf{p}} \cos^2\theta \tag{17.12}$$

where a factor 2 comes from the two polarizations. The restriction $p_x > 0$ means that $0 < \theta < \frac{\pi}{2}$. Going to spherical coordinates, we get a factor $\frac{1}{3}$ from the θ integration, and obtain

$$P = \frac{8\pi c}{3h^3} \int_0^\infty dp \frac{p^3}{e^{\beta c p} - 1} \tag{17.13}$$

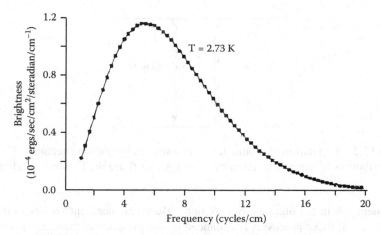

Figure 17.1 Observed frequency spectrum of cosmic background radiation. Dotted line is the Planck distribution with $T = 2.73$. (The unit "cm" denotes $\frac{1}{3} \times 10^{-10}$ s, the time for light to travel 1 cm.) The measurements are so accurate that there is no visible difference between theory and experiment over the whole frequency spectrum. (After Mather 1990.)

Comparison with Equation (17.7) leads to the relation

$$P = \frac{1}{3}\frac{U}{V} \tag{17.14}$$

This relation also holds for massive particles at temperatures so high that they move at close to light speed. It is to be compared with the relation $P = \frac{2}{3}U/V$ for a nonrelativistic gas of particles.

Our universe is filled uniformly with black-body radiation, as revealed in the measurements shown in Figure 17.1. The experimental points fit perfectly a Planck distribution of temperature 2.735 ± 0.05 K. Called the "cosmic background radiation," it is thought to be a relic of the Big Bang.

17.2 Bose Enhancement

We have seen in Section 14.13 that there is a statistical attraction between free bosons, even when there is no interaction potential. The difference between the Planck distribution and the Maxwell–Boltzmann distribution can be attributed to this attraction, as Einstein showed, in what has come to be known by the historical but clumsy name of "theory of A and B coefficients."

Consider a two-level atom in the wall of a black-body cavity, with level a at a higher energy than level b. Suppose the energy difference is $\hbar\omega$, so that the population of the two levels in equilibrium are maintained by emission or absorption of photons

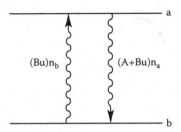

Figure 17.2 Population of atomic levels maintained by photon exchange. The photon distribution function is denoted by u, and A and B are the Einstein coefficients.

of frequency ω in the black-body radiation. "Bose enhancement" refers to the fact that the rate of these processes is enhanced by the presence of photons of the same frequency.

Let n_a and n_b denote respectively the number of atoms in level a and b. Einstein postulates the following rate equations:

$$\frac{dn_a}{dt} = -(A + Bu)n_a + Bun_b$$

$$\frac{dn_b}{dt} = (A + Bu)n_a - Bun_b \qquad (17.15)$$

where $u = u(\omega, T)$ is the energy density of photons per unit frequency. The gain and loss for level a are illustrated in Figure 17.2. Since atoms leaving a must go to b, and vice versa, we must have $d\,(n_a + n_b)\,/dt = 0$. The coefficient A is the rate of *spontaneous emission*, an intrinsic process independent of the environment. The coefficient B expresses Bose enhancement. It gives the rate of *stimulated emission* induced by the presence of other photons of the same frequency.

In equilibrium we must have $dn_a/dt = dn_b/dt = 0$, which requires

$$(A + Bu)n_a = Bun_b \qquad (17.16)$$

On the other hand, the ratio of the populations should be given by the Boltzmann factor

$$\frac{n_a}{n_b} = e^{-\beta\hbar\omega} \qquad (17.17)$$

Combining the last two equations, we can solve for u, and obtain

$$u(\omega, T) = \frac{A}{B}\frac{1}{e^{\beta\hbar\omega} - 1} \qquad (17.18)$$

Comparison with Equation (17.11) gives $A/B = \hbar\omega^3/\pi^2 c^3$. If we had omitted the stimulated emission term Bu on the left side of Equation (17.16), we would have obtained the Maxwell–Boltzmann distribution $u = (A/B)e^{-\beta\hbar\omega}$.

17.3 Phonons

A solid can be idealized as a crystal lattice of atoms, and the small-amplitude vibrations can be described in terms of normal modes, which are harmonic oscillators. The quanta of these normal modes are called *phonons*. They obey Bose statistics, and their number is not conserved. A phonon of momentum **p** has energy

$$\epsilon_{\mathbf{p}} = cp \tag{17.19}$$

where $p = |\mathbf{p}|$, c is the propagation velocity, and ω is the frequency. There are three modes of oscillation: one longitudinal mode corresponding to sound waves, and two transverse modes. We have taken the velocities of all three modes to be equal for simplicity.

The difference between a phonon and a photon, apart from the existence of a longitudinal mode, is that the frequency spectrum of a phonon has an upper cutoff, owing to the finite lattice spacing of the underlying crystal. Consider the simple example of beads on a string, as shown in Figure 17.3. We see that the half-wavelength of oscillations cannot be less than the lattice spacing.

Debye proposes the following model for the normal modes of a solid. Consider waves in a box of volume V, with periodic boundary conditions. The modes are plane wave labeled by a wave vector **k**. The number of modes in the element d^3k is

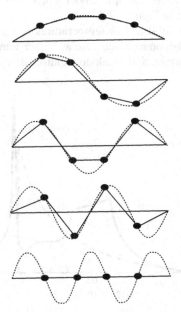

Figure 17.3 Collective oscillations of a finite number of particles have a minimum wavelength, hence a maximum cutoff frequency.

$Vd^3k/(2\pi)^3$. The frequency distribution function $f(\omega)$ is defined by

$$f(\omega)\,d\omega \equiv \frac{3V}{(2\pi)^3} 4\pi k^2 dk \qquad (17.20)$$

where $\omega = ck$ and the factor 3 takes into account the longitudinal and transverse modes. Thus,

$$f(\omega) = \frac{3\omega^2 V}{2\pi^2 c^3} \qquad (17.21)$$

The total number of modes of the system must be equal to $3N$, where N is the number of atoms in the lattice. The cutoff frequency ω_m, or *Debye frequency,* is defined by the requirement

$$\int_0^{\omega_m} f(\omega)\,d\omega = 3N \qquad (17.22)$$

which gives

$$\omega_m = (6\pi^2 n)^{1/3} \qquad (17.23)$$

where $n = N/V$ is the density of atoms. The characteristic energy $\hbar\omega_m$ defines the *Debye temperature* T_D:

$$k_B T_D = \hbar\omega_m \qquad (17.24)$$

Real solids have a more complicated frequency spectrum due to specifics of the crystalline structure, but Debye's simple model reproduces the quadratic behavior at low frequencies, as shown in Figure 17.4.

In the Debye model, a solid at low temperatures is represented by a collection of phonons. The picture becomes inaccurate at higher temperatures, when phonon interactions become important, and break down altogether when the solid melts.

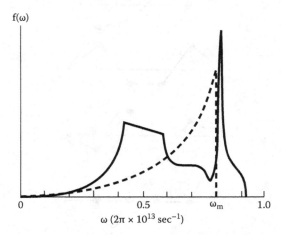

Figure 17.4   Solid curve is the fequency spectrum of normal modes of Al. (After Walker 1956.) Dashed curve is that of the Debye model.

17.4 Debye Specific Heat

In the Debye model the internal energy per atom is given by

$$\frac{U}{N} = \frac{1}{N}\sum_i \epsilon_i n_i = \frac{3V}{N(2\pi)^3}\int_0^{k_m} dk(4\pi k^2)\frac{\hbar\omega}{e^{\beta\hbar\omega}-1}$$

$$= \frac{12\pi\hbar}{n(2\pi)^3}\int_0^{\omega_m} d\omega\frac{\omega^3}{e^{\beta\hbar\omega}-1} \tag{17.25}$$

By changing the variable of integration to $t = \beta\hbar\omega$, and introducing the variable

$$u \equiv \frac{T_D}{T} \tag{17.26}$$

we can rewrite the energy per atom in the form

$$\frac{U}{N} = 3k_B T\, D(u) \tag{17.27}$$

where and $D(u)$ is the Debye function

$$D(u) = \frac{3}{u^3}\int_0^u dt\frac{t^3}{e^t - 1} \tag{17.28}$$

which has the asymptotic behaviors

$$D(u) \approx 1 - \frac{3}{8}u + \frac{1}{20}u^2 + \cdots \quad (u \ll 1)$$

$$D(u) \approx \frac{\pi^4}{5u^3} + O(e^{-u}) \quad (u \gg 1) \tag{17.29}$$

The specific heat in the Debye model is given by

$$\frac{C_V}{Nk} = 3D(u) + 3T\frac{dD(u)}{dT} \tag{17.30}$$

which is a universal function of T/T_D. The universal character is verified by experiments, as we can see from the graph in Figure 17.5. Near absolute zero, the specific behaves like T^3, which is a reflection of the linear dispersion law $\epsilon = cp$ for phonons. It deviates from this form at higher temperatures, when the effect of the cutoff becomes important.

The high-temperature asymptote $3Nk$ conforms to the *law of Dulong and Petit*. This value can be understood in terms of the equipartition of energy among $3N$ harmonic oscillators, each with energy $p^2 + q^2$ (in appropriate units), and thus contributing a term k to the heat capacity. Of course, the high-temperature limit here means that all degrees of freedom of the lattice are fully excited, but not so high that it melts—an effect that lies outside of this model.

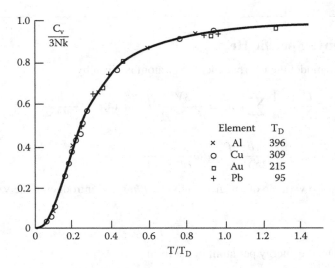

Figure 17.5 Universal curve for the specific heat in the Debye model. Data for different elements are fitted by adjusting the Debye temperature T_D. (After Wannier 1966.)

17.5 Electronic Specific Heat

There are electrons in the lattice of a solid, and they contribute to the specific heat of the solid. How important the contribution is depends on the temperature, and at room temperature it is completely negligible. As we can see from Figure 17.5, the Debye specific heat at $T = 300$ K for common metals is of order $3k_B$, but the electronic contribution is of order $(T/T_F)k_B$, with $T_F \sim 40000$ K. Thus the electronic contribution amounts to less than 1%. As the temperature decreases, however, the electronic contribution becomes more important. If we expand the specific heat in powers of T, then up to order T^3 it should have the form

$$\frac{C_V}{Nk} = c_1 \left(\frac{T}{T_F}\right) + c_2 \left(\frac{T}{T_F}\right)^2 + c_3 \left(\frac{T}{T_F}\right)^3 + c_4 \left(\frac{T}{T_D}\right)^3 + \cdots \qquad (17.31)$$

where the first three terms come from the electronic Fermi gas, and the last term comes from the Debye model. The c_n are numerical coefficients. Since $T_F \gg T_D$, we can neglect the two terms in the middle, and approximate the above by

$$\frac{C_V}{Nk} \approx aT + bT^3 \qquad (17.32)$$

Thus, at low temperatures, a plot of C_V/T against T^2 should give a straight line. This is indeed borne out by experiments, as shown in Figure 17.6. The density of electrons can be obtained from the data. (See Problem 17.10).

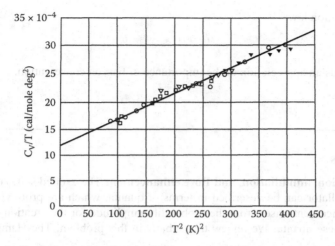

Figure 17.6 Data on specific heat of *KCl*. (After Keesom and Pearlman 1953.) The straightline is theory, which includes both electronic and lattice contributions.

17.6 Conservation of Particle Number

Why are some particle numbers conserved and others not? It depends entirely on their interactions. Phonons in a solid are not "real" particles, for we know that they are collective modes of the atoms in the solid. They are created and absorbed singly, because that's the way they interact with the atomic lattice. On the other hand, photons and electrons are "real" because we have no evidence otherwise.

Photons are created and absorbed singly by charged particles. Electrons can be created and destroyed only together with their antiparticle, the positron, by emission of a photon. Thus, the number of electrons minus the number of positrons is conserved. If there are no positrons present, or if the energy of 1 MeV required to create an electron-positron pair is not available, then the electron number is effectively conserved. Similarly, the number of baryons (of which the nucleon is the lightest example) minus the number of antibaryons is conserved. The pattern of fundamental interactions is such that bosons can be created or annihilated singly, while fermions are created or annihilated only in association with antifermions (Huang 1992).

Atoms are stable in the everyday world, because there are no antinucleons present. It requires at least 2 GeV to create a nucleon-antinucleon pair, and this greatly exceeds the thermal energy at room temperature. Were it not for the conservation of baryons, the bundle of energy locked up in the rest mass of the proton would prefer to assume other forms that give a larger entropy. The hydrogen atom would quickly disappear in a shower of mesons and gamma rays. We can use the Boltzmann factor to estimate the probability for finding a proton of rest mass M at room temperature, if there were no conservation law:

$$e^{-Mc^2/k_BT} \approx e^{-3\times 10^{10}} \tag{17.33}$$

This gives us an appreciation of the importance of baryon conservation.

Problems

17.1 Creation, annihilation, and Bose enhancement The excited states of the harmonic oscillator can be described in terms of quanta, which are prototype bosons. Bose enhancement arises from the probability amplitude for the creation or annihilation of these quanta. We review this subject in this problem. The Hamiltonian is given by

$$H = \frac{1}{2m}p^2 + \frac{m\omega^2}{2}q^2$$

where p, q are hermitian operators defined by the commutation relation $[p, q] = -i\hbar$.

(a) Make the transformation to *annihilation operator a* and *creation operator $a^\dagger$*:

$$p = \sqrt{\frac{\hbar m\omega}{2}}(a + a^\dagger)$$

$$q = i\sqrt{\frac{\hbar}{2m\omega}}(a - a^\dagger)$$

Show that

$$H = \hbar\omega\left(a^\dagger a + \frac{1}{2}\right)$$

$$[a, a^\dagger] = 1$$

(b) Show that the eigenvalues of $a^\dagger a$ are integers $n = 0, 1, 2, \ldots$.

(c) Let $|n\rangle$ be an eigenstate of $a^\dagger a$ belonging to the eigenvalue n. Show that

$$a|n\rangle = \sqrt{n}|n-1\rangle$$

$$a^\dagger|n\rangle = \sqrt{n+1}|n+1\rangle$$

Thus, a annihilates one quantum, and $a^\dagger$ a creates one quantum. The factors $\sqrt{n}$, $\sqrt{n+1}$ give rise to Bose enhancement.

17.2 A black body is an idealized object that absorbs all radiation falling upon it, and emits radiation according to Stefan's law. A star is maintained by internal nuclear burning at an absolute temperature T. It is surrounded by a dust cloud heated by radiation from the star. Treating both the star and the dust cloud as black bodies, show that in radiative equilibrium

(a) the dust cloud reduces the radiative flux from the star to the outside world by half;

(b) the temperature of the dust cloud is $2^{-4}T$.

17.3 Assume that the Earth's temperature is maintained through heating by the Sun. Treating both the Sun and the Earth as black bodies, show that the ratios of the Earth's temperature to the Sun's is given by

$$\frac{T_{\text{Earth}}}{T_{\text{Sun}}} = \sqrt{\frac{R_{\text{Sun}}}{2L}}$$

where R_{Sun} is the radius of the Sun, and L is the distance between the Sun and the Earth. We have ignored the heating of the Earth due to radioactivity in the interior. (See Problem 7.8.)

17.4 Suppose a house exposed to heat radiation has a reflection coefficient r, that is, it reflects a fraction r, and accepts a fraction $1 - r$, of incident radiation. Assume that it radiates heat like a black body. Let the temperature inside the house be T, and that outside be T_0. Show that under equilibrium conditions

$$\frac{T}{T_0} = \left(\frac{1-r}{1-2r}\right)^{1/4}$$

17.5 Show that the entropy of a photon gas is given by $S = \frac{4}{3}U/T$, hence the entropy density is

$$\frac{S}{V} = \frac{4}{3}\sigma T^3$$

where $\sigma = \pi^2 k^4 / [15(\hbar c)^3]$ is Stefan's constant.

17.6 The background cosmic radiation has a Planck distribution temperature with temperature 2.73 K, as shown in Figure 17.1.

(a) What is the photon density in the universe?

(b) What is the entropy per photon?

(c) Suppose the universe expands adiabatically. What would the temperature be when the volume of the universe doubles?

17.7 Einstein model of solid: The Einstein model is a cruder version of the Debye model. It assumes that all phonons have the same frequency ω_0.

(a) Give the internal energy of this model.

(b) Find the heat capacity. How does it behave near absolute zero?

(c) Show that the free energy is given by

$$A = 3Nk_B T \ln(1 - e^{-\beta\hbar\omega_0})$$

17.8 A solid exists in equilibrium with its vapor, which is treated as a classical ideal gas. Assume that the solid has a binding energy ϵ per atom. Describe the phonons using the Einstein model.

(a) Give the free energy for the total system consisting of solid and vapor. (For the classical ideal gas, you must use correct Boltzmann counting.)

(b) What is the condition for equilibrium between vapor and solid?

(c) Find the vapor pressure $P(T)$.

17.9 A model of a solid taking into account the structural energy of the lattice postulates a free energy of the form

$$A(V, T) = \phi(V) + A_{\text{phonon}}$$

where $\phi(V)$ is a structural energy, and A_{phonon} is the contribution from phonons. At absolute zero, there are no phonons, and the volume of the solid is determined by minimizing $\phi(V)$. Near the minimum we may write

$$\phi(V) = -\phi_0 + \frac{K}{2}(V - V_0)^2$$

At higher temperatures, there is a shift in the minimum due to the fact that the phonon frequency increases when the lattice spacing decreases:

$$\frac{d\omega(k)}{\omega(k)} = -\gamma \frac{dV}{V}$$

where γ is known as Gruneisen's constant.

(a) Neglect the vapor pressure and take the pressure of the solid to be zero. Find the equilibrium condition using the Einstein model for the phonon contribution.

(b) Find the equilibrium volume of the solid as a function of temperature, and calculate the coefficient of thermal expansion.

17.10 In the specific heat formula [Equation (17.32)] for solids $C_V/Nk = aT + bT^3$, the first term represents the contribution from electrons in the solid, and the second from phonons.

(a) Give a in terms of the electron density n,

(b) Give b in terms of the Debye temperature T_D.

(c) Obtain the numerical values of a and b from the graph in Figure 10.6 for KCl, and find n and T_D.

17.11 Surface waves on liquid helium has a dispersion relation given by

$$\epsilon(k) = \hbar \sqrt{\frac{\sigma k^3}{\rho}}$$

where k is the wave number of the surface wave, σ the surface tension, and ρ the mass density of the liquid. Treating the excitations as bosons with no number conservation, find the internal energy per unit area as a function of temperature.

References

Huang, K., *Quarks, Leptons, and Gauge Fields*, 2nd ed., World Scientific Publishing, Singapore, 1992.

Keesom, P.H. and N. Pearlman, *Phys. Rev.*, **91:**1354 (1953).

Mather, J.C. et al., *Ap. J.*, **354:**L37 (1990).

Walker, C.B., *Phys. Rev.*, **103:**547 (1956).

Wannier, G.H., *Statistical Physics*, Wiley, New York, 1966.

References

Chapter 18

Bose–Einstein Condensation

18.1 Macroscopic Occupation

Consider a gas of bosons whose total number is conserved. The equation for the fugacity is

$$n\lambda^3 = g_{3/2}(z) \tag{18.1}$$

and a qualitative plot of the function $g_{3/2}$ is shown in Figure 18.1. The function has infinite slope at $z = 1$, since

$$z\frac{d}{dz}g_{3/2}(z) = g_{1/2}(z) = \sum_{\ell=1}^{\infty} \frac{z^\ell}{\sqrt{\ell}} \tag{18.2}$$

and the series diverges at $z = 1$. It is thus bounded by the value

$$g_{3/2}(1) = \sum_{\ell=1}^{\infty} \ell^{-3/2} = \varsigma(3/2) = 2.612\cdots \tag{18.3}$$

where $\zeta(z)$ is the Riemann zeta function. There is no solution to Equation (18.1) unless

$$n\lambda^3 \leq g_{3/2}(1) \tag{18.4}$$

This condition can be violated by increasing the density, or lowering the temperature. What then would be the value of z ?

To answer this question, recall that in deriving Equation (18.1) in Section 14.12, we took the infinite-volume limit, and replaced the sum over states by an integration. This assigns zero weight to the state of zero momentum, for the volume element $4\pi p^2 dp$ vanishes at $p = 0$. In the formula (Equation [18.1]), n is therefore the density of particles with nonzero momentum, and this is the quantity bounded by Equation (18.4). When this bound is exceeded, the excess particles cannot disappear, as photons could, because their number is conserved. They are forced to go into the zero-momentum state, and this single quantum state becomes *macroscopically occupied*, that is, occupied by a finite fraction of the particles. This phenomenon is called *Bose–Einstein condensation*.

251

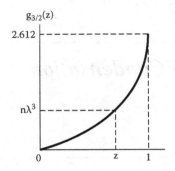

Figure 18.1 This function refers to particles with nonzero momentum, and is bounded by $g_{3/2}(1) = 2.612$. When $n\lambda^3$ exceeds this value, particles must go into the zero-momentum state, leading to Bose–Einstein condensation.

The fugacity z is stuck at 1 during the Bose–Einstein condensation. Thus,

$$z = \begin{cases} \text{Root of } n\lambda^3 = g_{3/2}(z) & \text{if } n\lambda^3 < g_{3/2}(1) \quad \text{(gas phase)} \\ 1 & \text{if } n\lambda^3 \geq g_{3/2}(1) \quad \text{(condensed phase)} \end{cases} \tag{18.5}$$

The critical value $n\lambda^3 = g_{3/2}(1)$ corresponds to

$$n \left(\frac{2\pi\hbar^2}{mk_B T} \right)^{3/2} = \varsigma(3/2) \tag{18.6}$$

This gives the boundary between the gas phase and condensed phase, as shown the n-T diagram of Figure 18.2. At fixed density n, the critical line defines a critical temperature T_c :

$$k_B T_c = \frac{2\pi\hbar^2}{m} \left[\frac{n}{\varsigma(3/2)} \right]^{2/3} \tag{18.7}$$

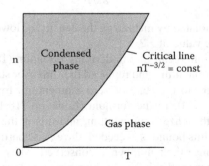

Figure 18.2 Phase diagram of Bose–Einstein condensation in the density-temperature plane.

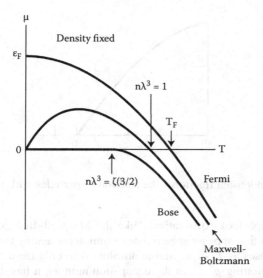

Figure 18.3 Qualitative behavior of the chemical potential for the Bose, Fermi, and the Maxwell–Boltzmann gas.

At a given temperature T, it defines a critical density n_c:

$$n_c = \varsigma(3/2) \left(\frac{mk_B T}{2\pi\hbar^2} \right)^{3/2} \tag{18.8}$$

The chemical potential $\mu = k_B T \ln z$ is zero in the condensed phase. Its qualitative behavior in comparison with the Fermi gas and classical gas is shown in Figure 18.3.

18.2 The Condensate

The Bose–Einstein condensate consists of the particles in the zero-momentum state. Their number n_0 is determined through the relation

$$N = n_0 + \frac{V}{\lambda^3} g_{3/2}(1) \tag{18.9}$$

Thus,

$$\frac{n_0}{N} = 1 - \frac{g_{3/2}(1)}{n\lambda^3} = 1 - \left(\frac{T}{T_c} \right)^{3/2} \tag{18.10}$$

This is illustrated in Figure 18.4.

The momentum distribution $p^2 n_p$ is shown in Figure 18.5 for both $T > T_c$ and $T < T_c$. The area under the curve gives the density of particles $4\pi h^{-3} \int_0^\infty dp\, p^2 n_p$.

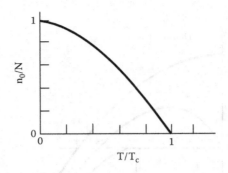

Figure 18.4 Condensate fraction—the fraction of particles with momentum zero.

Above T_c the graph looks qualitatively like the Maxwell–Boltzmann distribution. Below T_c, a δ function appears at zero momentum, representing the condensate, and its strength increases as $T \rightarrow 0$, and at absolute zero only the δ-function remains. Because we are plotting p^2 times the occupation number, it has the rather singular form $p^{-2}\delta(p)$.

18.3   Equation of State

The equation of state as a function of fugacity was derived in Section 14.12:

$$\frac{P}{k_B T} = \frac{g_{5/2}(z)}{\lambda^3} \qquad (18.11)$$

This is valid in both the gas and condensed phase, because particles with zero-momentum do not contribute to the pressure. Since $z = 1$ in the condensed phase,

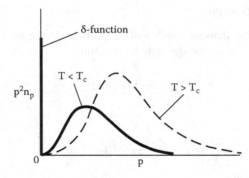

Figure 18.5 Area under the curve is proportional to the density of particles. Below the transition temperature, a δ-function appears at zero momentum, representing the condensate contribution.

the pressure becomes independent of density:

$$\frac{P}{k_B T} = \frac{g_{5/2}(1)}{\lambda^3} \tag{18.12}$$

where $g_{5/2}(1) = 3.413 \cdots$.

In the gas phase we have to solve for z from Equation (18.1). At low densities, which correspond to small z, the system approaches a classical ideal gas. At higher densities, we have to obtain the solution through numerical methods. The qualitative isotherms are shown in the PV diagram in Figure 18.6. The Bose–Einstein condensation appears as a first-order phase transition. The vapor pressure is independent of the density:

$$P = C_0 (k_B T)^{5/2}$$

$$C_0 = \left(\frac{m}{2\pi\hbar^2} \right)^{3/2} g_{5/2}(1) \tag{18.13}$$

It can be verified that the vapor pressure satisfies the Clapeyron equation.

The horizontal part of an isotherm in Figure 18.6 represents a mixture of the gas phase and the pure condensate, and the latter occupies zero volume because there is no inter-particle interaction. In a real Bose gas, which has nonzero compressibility, the Bose–Einstein condensation would be a second-order, instead of a first-order phase transition. This has been observed indirectly in liquid helium, which we shall describe later, and more explicitly in dilute atomic gases trapped in an external potential, which will be discussed in Chapter 20.

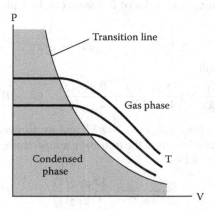

Figure 18.6 Qualitative isotherms of the ideal Bose gas. The Bose–Einstein condensation shows up as a first-order phase transition.

18.4   Specific Heat

The internal energy is given by

$$U = \frac{3}{2}PV \tag{18.14}$$

The heat capacity per particle at constant volume is therefore

$$\frac{C_V}{Nk_B} = \frac{3}{2nk_B}\left(\frac{\partial P}{\partial T}\right)_V \tag{18.15}$$

We shall calculate this separately in the condensed and gas phases. In the condensed phase $P = C_0 T^{5/2}$. Therefore

$$\frac{C_V}{Nk_B} = \frac{15C_0}{4nk_B}(k_B T)^{3/2} \quad (T < T_c) \tag{18.16}$$

The behavior near absolute zero should be compared with other types of behavior we know:

- $T^{3/2}$, for conserved bosons with $\epsilon \propto p^2$;
- T, for conserved fermions with $\epsilon \propto p^2$;
- T^3, for nonconserved bosons with $\epsilon \propto p$.

In the gas phase we have

$$\left(\frac{\partial P}{\partial T}\right)_V = \left(\frac{\partial}{\partial T}\frac{k_B T}{\lambda^3}\right)_V g_{5/2}(z) + \frac{k_B T}{\lambda^3}\left(\frac{\partial}{\partial T}g_{5/2}(z)\right)_V$$

$$= \frac{5k_B}{2\lambda^3}g_{3/2}(z) + \frac{k_B T}{\lambda^3}g_{3/2}(z)\left(\frac{1}{z}\frac{\partial z}{\partial T}\right)_V \tag{18.17}$$

where z must be regarded as a function of T and V: The derivative $(\partial z/\partial T)_V$ can be obtained by differentiating both sides of Equation (18.1), with the result

$$\frac{1}{z}\left(\frac{\partial z}{\partial T}\right)_V = -\frac{3n\lambda^3}{2T}\frac{1}{g_{1/2}(z)} \tag{18.18}$$

Thus we obtain the result

$$\frac{C_V}{Nk_B} = \frac{15}{4}\frac{g_{5/2}(z)}{n\lambda^3} - \frac{9}{4}\frac{g_{3/2}(z)}{g_{1/2}(z)} \quad (T > T_c) \tag{18.19}$$

In the high-temperature limit $g_n(z) \approx z \approx n\lambda^3$, and we recover the classical limit $\frac{3}{2}$.

The specific heat is continuous at the transition temperature, as we can verify from Equations (18.16) and (18.19):

$$\frac{C_V}{Nk_B} \xrightarrow[T \to T_c]{} \frac{15}{4}\frac{g_{5/2}(1)}{g_{3/2}(1)} \tag{18.20}$$

However, the slope is discontinuous. (See Problem 18.4.) A qualitative plot of the specific heat is shown in Figure 18.7.

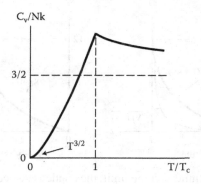

Figure 18.7

18.5 How a Phase is Formed

The Bose–Einstein condensation of an ideal gas is one the few examples of a phase transition we can completely describe mathematically. It is interesting to see how the equation of state develops discontinuous behavior in the infinite-volume limit.

To examine the onset of the Bose–Einstein condensation, let us go back to the condition for z at finite volume:

$$\frac{N}{V} = \frac{1}{V} \sum_{\mathbf{p}} \frac{1}{z^{-1} e^{\beta p^2} - 1} \tag{18.21}$$

In taking the limit $V \to \infty$ we make the replacement

$$\frac{1}{V} \sum_{\mathbf{p}} \to \int \frac{d^3 p}{h^3} \tag{18.22}$$

which assigns zero measure to the state $\mathbf{p} = 0$. In doing so, we have ignored the density of the particles in the zero-momentum state:

$$\frac{n_0}{V} = \frac{1}{V} \frac{z}{1 - z} \tag{18.23}$$

We have a case of nonuniform convergence here, for this quantity approaches different values depending the order in which we take the limits $V \to \infty$ and $z \to 1$:

$$\lim_{z \to 1} \lim_{V \to \infty} \frac{n_0}{V} = 0$$

$$\lim_{V \to \infty} \lim_{z \to 1} \frac{n_0}{V} = \infty \tag{18.24}$$

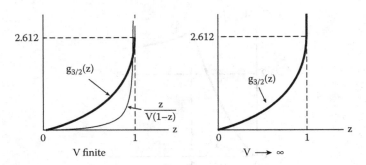

Figure 18.8 In the infinite-volume limit, the condensate occupation number effectively adds a straight vertical rise to $g_{3/2}$ at $z = 1$.

Let us separate out this term in Equation (18.21), and replace the rest by the large V limit:

$$\frac{N}{V} = \frac{1}{V}\frac{z}{1-z} + \int \frac{d^3 p}{h^3}\frac{1}{z^{-1}e^{\beta p^2}-1}$$

$$= \frac{1}{V}\frac{z}{1-z} + \frac{1}{\lambda^3}g_{3/2}(z) \qquad (18.25)$$

The two terms on the right side are shown in Figure 18.8, for finite V and for $V \to \infty$. In the limit $V \to \infty$, the term $V^{-1}z/(1-z)$ is zero for $z < 1$, and indeterminate at $z = 1$. Thus, the density of zero-momentum particles n_0/V is zero for $z < 1$, but assumes whatever value is required to satisfy Equation (18.21) at $z = 1$.

The fugacity as a function of $n\lambda^3$ is shown in Figure 18.9. It is a continuous function for finite V, however large. In the limit $V \to \infty$, however, it approaches different functions for $n\lambda^3 > \zeta(3/2)$, and $n\lambda^3 < \zeta(3/2)$. In particular, it is a constant function $z = 1$ in the latter region.

In summary, the equation of state is a regular function of n and T for finite volume, however large. However, in the limit $V \to \infty$ it can approach different limiting forms for different values of n and T. Although the idealized discontinuous phase transition occurs only in the thermodynamic limit, macroscopic systems are sufficiently close to that limit as to make the idealization useful.

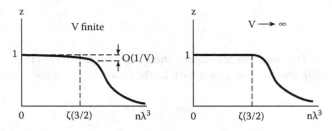

Figure 18.9 The fugacity approaches different limiting functions for different ranges of $n\lambda^3$, in the infinite-volume limit.

18.6   Liquid Helium

The atom ^{4}He, the most abundant isotope of helium, is a boson. We might expect Bose–Einstein condensation to occur in bulk helium, which is a liquid at low temperatures. Indeed, there is a second-order phase transition in liquid helium that can be identified with this phase transition, which is called the "λ-transition" after the shape of the specific heat shown in Figure 18.10. Supporting the identification is the fact that below T_c the liquid is a "superfluid" that manifests quantum behavior on a macroscopic scale, such as the absence of viscosity. Because of interatomic interactions, the specific heat of liquid helium behaves like T^3, instead of $T^{3/2}$ as in the ideal Bose gas, indicating the existence of phonon excitations. The interactions apparently change the order of the transition changed from first to second. Properties at the transition point are given by

$$T_c = 2.172 \text{ K}$$

$$n_c = 2.16 \times 10^{22} \text{cm}^{-3}$$

$$v_c = 46.2 \text{ A}^3 \text{ per atom} \tag{18.26}$$

Substituting n_c into Equation (18.7) yields $T_c \approx 3.15$ K, which is of the right order of magnitude.

Low-energy excitations in liquid helium can be detected through neutron scattering. The experimental dispersion relation $\epsilon(k)$, which gives the energy of the quasiparticle as a function of wave number, is shown in Figure 18.11, which shows the phonon branch near $k = 0$,

$$\epsilon_{\text{phonon}} = \hbar c k$$

$$c = 239 \pm 5 \text{ m/s} \tag{18.27}$$

where c is the sound velocity.

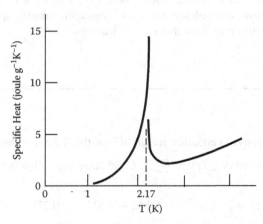

Figure 18.10   λ-transition at $T_c = 2.172$ K goes between two liquid phases, both in equilibrium with a gas phase. The specific heat is measured along the vapor pressure curve. (After Hill and Lounasmaa 1957.)

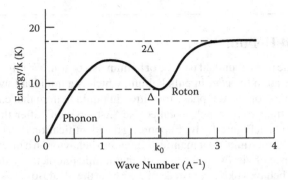

Figure 18.11 Dispersion curve of quasiparticles in liquid ^{4}He obtained through neutron scattering. [Donelly et al., 1981, ($0 < k < .2.5$ A^{-1}); Glyde et al. 1998, ($k > 2.5$ A^{-1}).] The data points are very dense, with errors about the width of the line plotted.

At higher values of k, the dispersion curve goes through a minimum, in the neighborhood of which the excitation behaves like a particle named the "roton," with the properties

$$\epsilon_{\text{roton}} = \Delta + \frac{\hbar^2(k - k_0)^2}{2\sigma}$$

$$\Delta = 8.65 \pm 0.04 \text{ K}$$

$$k_0 = 1.92 \pm 0.01 \text{ A}^{-1}$$

$$\sigma = (0.16 \pm 0.01)m \qquad (18.28)$$

where m is the mass of a helium atom. These excitations are quite different from those in the ideal Bose gas with spectrum $\hbar^2 k^2/2m$, and make the specific heat at low temperatures very different from that of the ideal gas.

Problems

18.1 Verify the Clapeyron equation for dP/dT for the Bose–Einstein condensation.

18.2 The Bose functions $g_n(z) = \sum_{\ell=1}^{\infty} \ell^{-n} z^\ell$ have the following expansions near $z = 1$:

$$g_{5/2}(z) = 2.363\, v^{3/2} + 1.342 - 2.612\, v - 0.730\, v^2 + \cdots$$

$$g_{3/2}(z) = 2.612 - 3.455\, v^{1/2} + 1.460\, v + \cdots$$

where $v = -\ln z$. They are related through $g_{3/2} = -dg_{5/2}/dv$.

Find the fugacity z of an ideal Bose gas of density n, as a function of the temperature T in the neighborhood of $T = T_c$. For $T < T_c$ we have of course $z = 1$. For $T \geq T_c$, obtain the first few terms of a power-series expansion in $T - T_c$.

18.3 Show that the equation of state of the ideal Bose gas in the gas phase has the virial expansion

$$\frac{P}{nk_B T} = 1 + a_2(n\lambda^3) + a_3(n\lambda^3)^2 + \cdots$$

where

$$a_2 = -\frac{1}{4\sqrt{2}}$$

$$a_3 = \frac{1}{8} - \frac{2}{9\sqrt{3}}$$

18.4 Show that the slope of the heat capacity of an ideal Bose gas has a discontinuity at $T = T_c$ given by

$$\left(\frac{\partial C_V}{\partial T}\right)_{T \to T_c^+} - \left(\frac{\partial C_V}{\partial T}\right)_{T \to T_c^-} = 3.66 \frac{Nk}{T_c}$$

Hint: Calculate the internal energy via $U = 3PV$.

18.5 Suppose the particle spectrum has a gap $\Delta > 0$:

$$\epsilon(k) = \begin{cases} -\Delta & (k = 0) \\ \hbar^2 k^2/2m & (k > 0) \end{cases}$$

Show how this would modify the Bose–Einstein condensation.

(a) From the equation determining the fugacity, show that the condensation happens when $z = \exp(-\beta\Delta)$.

(b) Assuming $\Delta/k_B T_0 \ll 1$, where T_0 is the transition temperature for $\Delta = 0$, find the shift in transition temperature $T_c - T_0$. Use the expansion of $g_{3/2}(z)$ near $z = 1$, given in Problem 18.2.

18.6 The heat capacity of liquid ^{4}He near absolute zero should be dominated by contributions from phonons and rotons.

(a) Calculate the contribution C_{phonon} from a gas of nonconserved bosons, with dispersion relation

$$\epsilon(k) = \hbar c k$$

(b) Calculate the contribution C_{roton} from a gas of nonconserved bosons with dispersion relation

$$\epsilon(k) = \Delta + \frac{\hbar^2(k - k_0)^2}{2\sigma}$$

18.7 Compare $C = C_{phonon} + C_{roton}$ to the specific heat data given in the following table. The phonon and roton data from scattering experiments are given in Equations (18.27) and (18.28). The number density and mass density of liquid ^{4}He are respectively $n = 2.16 \times 10^{22}$ cm^{-3}, and $\rho = 0.144$ g/cm^3.

Temp (K)	SpHeat (J/g-deg)
0.60	0.0051
0.65	0.0068
0.70	0.0098
0.75	0.0146
0.80	0.0222
0.85	0.0343
0.90	0.0510
0.95	0.0743
1.00	0.1042

18.8 Consider a box divided into two compartments. One compartment contains an almost degenerate ideal Fermi gas of atoms with mass m_1 and density n_1. The other compartment contains an ideal Bose gas of atoms of mass m_2 and density n_2 below the transition temperature for Bose–Einstein condensation. The dividing wall can slide without friction, and conducts heat, so the two gases come equilibrium with the same pressure and temperature.

(a) Show that the pressure of the Fermi gas has the form

$$P_1 = c_1 \left(\frac{\hbar^2}{m_1} \right) n_1^{5/2}$$

where c_1 is a numerical constant.

(b) Show that the vapor pressure of the Bose gas has the form

$$P_2 = c_2 \left(\frac{m_2}{\hbar} \right)^{3/2} (k_B T)^{5/2}$$

where c_2 is a numerical constant.

(c) Give the condition on the temperature for the Fermi gas to be approximately degenerate.

(d) Give the condition for the pressure to equalize.

(e) Find the condition on m_1/m_2 in order that (c) and (d) are both satisfied.

18.9 By repeating the argument for Bose–Einstein condensation in 3D, show that in 2D the transition temperature approaches zero in the limit of an infinitely large system.

18.10 If the mechanism for photon absorption or emission can be neglected, as may happen in some cosmological settings, the number of photons would be conserved.

Can a photon gas undergo Bose–Einstein condensation under these circumstances? If so, give the critical photon density at temperature T.

18.11 Consensate fluctuations In the region of Bose–Einstein transition, where $z = 1$, the mean-square fluctuation of the occupation number is, according to Equation (15.34),

$$\langle n_{\mathbf{k}}^2 \rangle - \langle n_{\mathbf{k}} \rangle^2 = \frac{e^{-\beta\epsilon}}{(1 - e^{-\beta\epsilon})^2}$$

where $\epsilon = \hbar^2 k^2 / 2m$. This diverges for the condensate, since $\mathbf{k} = 0$. But this is for an infinite volume. Consider a finite but large cubical box of volume $V = L^3$. Show that, as $L \to \infty$, the mean-square fluctuation for n_0 diverges like $L^4 = V^{4/3}$.

Solution:

The boundary condition should not matter in the infinite-volume limit. With a wave function that vanishes on the boundary, the lowest single-particle state has deBroglie wavelength $\lambda = 2L$, and its energy is

$$\epsilon = \frac{\hbar^2}{2m} \left(\frac{2\pi}{\lambda} \right)^2 = \frac{\hbar^2}{2m} \left(\frac{\pi}{L} \right)^2$$

Thus the mean-square fluctuation is

$$\frac{e^{-\beta\epsilon}}{(1 - e^{-\beta\epsilon})^2} \approx (\beta\epsilon)^{-2} = \left(\frac{2mk_B T}{\pi^2 \hbar^2} \right)^2 L^4$$

References

Donelly, R.J. et al., *J. Low Temp. Phys.*, **44**:471 (1981).

Glyde, H.R. et al., *Europhys. Lett.*, **43**:422 (1998).

Hill, R.W. and O.V. Lounasmaa, *Phil. Mag.*, **2**:143 (1957).

Chapter 19

The Order Parameter

19.1 The Essence of Phase Transitions

In a phase transition, a system changes from a less "ordered" state to a more ordered one, or vice versa. The precise meaning of "order" depends on the system, but the general idea of a change in order applies to all. Take the ferromagnetic transition, for example. The directions of atomic magnetic moments change from random orientation to alignment, as illustrated in Figure 19.1. We can describe order through the average magnetic moment $\phi(x)$, which can be obtained by averaging the moment vectors in a small neighborhood about the point x. If the orientations are random we would get $\phi(x) = 0$. Otherwise we get a varying field, and when uniform magnetization is established we will get a constant field over all space. This is the prototype of an *order parameter.*

In general, the order parameter is a field over a D-dimensional spaces, with any number of components. For physical systems $D = 1, 2, 3$; but for theoretical purposes we admit other values, including noninteger ones. In the ferromagnetic case it is a vector with three real components, but in general it may be a complex number, or anything the system calls for. We shall use the notation $\phi(x)$ to denote the generic case.

The field $\phi(x)$ actually contains more details than needed to explain the usual sort of experimental data. In the ferromagnetic transition, for example, one measures the total magnetic moment, that is, the integral $\int dx \phi(x)$ over the entire sample. But the local magnetization density $\phi(x)$ supplies information concerning fluctuations, and gives a deeper view of the phase transition.

A microscopic description of the system involves many other degrees of freedom beyond the order parameter. In a ferromagnetic system, the basic Hamiltonian involves all atomic details, such as electronic states, interatomic interactions, etc. Calculation of the partition function using the Hamiltonian would encounter insurmountable mathematical difficulty. Direct numerical computations are beyond the reach of present-day computers by orders of magnitude. An alternative is to try to build a phenomenological theory in terms of the order parameter alone, hiding all other degrees of freedom in terms of coefficients in the theory. This is the Ginsburg–Landau approach that will occupy us here.

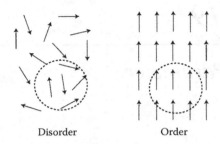

Disorder   Order

Figure 19.1 The average local magnetization, obtained by averaging the atomic magnetic moments in a neighborhood of x (indicated by the dotted circle) is a prototype of the order parameter. The dotted circle should contain a very large number of atoms.

19.2 Ginsburg–Landau Theory

Consider, for simplicity, a one-component order parameter, a scalar field $\phi(x)$ over a space of any number of dimensions. The state of the system is specified by $\phi(x)$, and the statistical ensemble describing thermal properties of the system is a set of such fields, with assigned statistical weights given by the Boltzmann factor

$$e^{-\beta E[\phi]} \tag{19.1}$$

where $\beta = (k_B T)^{-1}$. The functional $E[\phi]$, called the *Landau free energy,* specifies the system (Historically, such a functional was first used in van der Waals 1893).

The Ginsburg–Landau theory is based on the postulate

$$E[\phi] = \int dx \left[\frac{\varepsilon^2}{2} |\nabla\phi(x)|^2 + W(\phi(x)) - h(x)\phi(x) \right] \tag{19.2}$$

The term containing $|\nabla\phi|^2$ is analogous to a kinetic energy. It imposes an energy cost for a spatial gradient, thus creating a tendency towards uniformity. The constant ε is a "stiffness coefficient." One can always remove ε by recalling, that is, absorbing it into $\phi(x)$.

The quantity $W(\phi)$ is analogous to a potential. We can expand the potential in powers of ϕ:

$$W(\phi(x)) = g_2\phi^2(x) + g_3\phi^3(x) + g_4\phi^4(x) + \cdots \tag{19.3}$$

where the coupling constants g_n are the phenomenological parameters of the theory, and may depend on the temperature. We did not include a linear term above, for it was separately displayed in Equation (19.2) in the form $h(x)\phi(x)$. Here, $h(x)$ represents an external field coupled linearly to the order parameter, such as a local magnetic field in a ferromagnetic system. This term is singled out, because it serves a very convenient mathematical purpose, as we see later.

A sum over states of the system is a sum over all permissible forms of $\phi(x)$—a functional integral, denoted by the notation $\int D\phi$. The partition function is then

$$Q[h] = \int D\phi \, e^{-\beta E[\phi]} \tag{19.4}$$

which is a functional of the external field h. The field $\phi_0(x)$ that minimizes $E[\phi]$ has dominant weight, and describes thermal equilibrium. The other ϕ's correspond to thermal fluctuations.

The free energy is a functional of the external field, given by

$$A[h] = -k_B T \ln Q[h] \tag{19.5}$$

The ensemble average of a quantity O is given by

$$\langle O \rangle = \frac{\int D\phi \, O e^{-\beta E[\phi]}}{\int D\phi \, e^{-\beta E[\phi]}} \tag{19.6}$$

In analogy with the total magnetization, the total order is defined as

$$M = \left\langle \int d^D x \, \phi(x) \right\rangle \tag{19.7}$$

This can be obtained from the free energy by the formula

$$M = -\frac{\partial A}{\partial h} \tag{19.8}$$

where h is a uniform external field. The susceptibility is defined by

$$\chi = \frac{1}{V} \frac{\partial M}{\partial h} \tag{19.9}$$

where V is the total volume of the system. The heat capacity at constant external field is given by

$$C = -T^2 \frac{\partial^2 A}{\partial T^2} \tag{19.10}$$

19.3 Relation to Microscopic Theory

The whole point of the Ginsburg–Landau theory is to avoid dealing with the underlying microscopic theory. But how is it mathematically related to it? Let us consider a ferromagnetic system as an example.

The microscopic structure is defined by a quantum-mechanical N-atom Hamiltonian, in which atoms have intrinsic magnetic moments. Let the eigenstate of the

N-body system be labeled by n, with energy eigenvalue E_n. Suppose $\mu_n(x)$ is the average magnetic moment density at the point x, in the nth eigenstate.[1]

The Landau free energy $E[\phi]$ is defined by gathering up all the quantum states with the same functional form of μ_n, through the formula

$$e^{-\beta E[\phi]} \equiv \sum_n \delta[\phi - \mu_n] e^{-\beta E_n} \qquad (19.11)$$

where $\delta[\phi - \mu_n]$ is the functional δ-function defined through the properties

$$\delta[\phi - \mu_n] = 0 \quad (\text{unless } \phi = \mu_n) \qquad (19.12)$$

$$\int D\phi \, \delta[\phi - \mu_n] = 1$$

The partition function (Equation [19.4]) then gives

$$Q = \int D\phi \, e^{-\beta E[\phi]} = \sum_n \int D\phi \, \delta[\phi - \mu_n] e^{-\beta E_n} = \sum_n e^{-\beta E_n} \qquad (19.13)$$

It is equal to the microscopic partition function.

Note that the microscopic degrees of freedom are "hidden," but not ignored. We have not changed the underlying theory, but merely expressed it in a different form. The hidden degrees of freedom affect the parameters of $E[\phi]$, and their entropy has been incorporated into $E[\phi]$, and that's why the latter is called a "free energy."

19.4   Functional Integration and Differentiation

We now explain how functional operations can be carried out in practice.

The functional integration may be performed as follows. First, replace the continuous space x by a discrete lattice of points $\{x_1, x_2, \dots \}$. As a shorthand let $\phi_i = \phi(x_i)$. A functional integral over ϕ can be approximated by the multiple integral over the independent values $\{\phi_1, \phi_2, \dots \}$:

$$\int D\phi = \int_{-\infty}^{\infty} d\phi_1 \int_{-\infty}^{\infty} d\phi_2 \cdots \qquad (19.14)$$

This is illustrated in Figure 19.2. We approach the continuum limit by making the lattice spacing smaller and smaller with respect to some fixed scale. In the limit, the

---

[1]In principle, the magnetic moment density is calculated as follows. Let $\psi(x)$ denote the quantized field operator describing the atomic system, and $|n\rangle$ the eigenstate. Then

$$\mu_n = \mu_0 \overline{\langle n | \psi^*(x) \, \sigma \, \psi(x) | n \rangle}$$

where σ is the spin operator, μ_0 the magnetic moment, and the overhead bar denotes spatial average about the point x. Vector indices have been omitted.

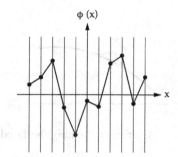

Figure 19.2 The functional integral over ϕ can be approximated by integrating over the values of $\phi(x_i)$ independently over a discrete set $\{x_i\}$. The continuum limit is taken after physical quantities are calculated with this method.

functional integration includes continuous as well as discontinuous functions. In fact, the vast majority of the functions are discontinuous, but because of their large kinetic contribution, they give small contributions to the partition function. The continuum limit of the functional integral itself may be ambiguous, but the limit of the ensemble average is usually well-defined.

Another way to define a function integration is to specify ϕ through its Fourier components

$$\phi_k = \int d^D x \phi(x) \, e^{ik \cdot x} \tag{19.15}$$

where x and k are D-dimensional vectors. When the system is enclosed in a large but finite box, with periodic boundary conditions, the allowed values of k constitute a discrete set. The functional integral can again be defined as Equation (19.14), only $d\phi_k$ now refers to an infinitesimal change in the Fourier component. The continuum limit can be taken after ensemble averages are computed.

To define the functional derivative, let us remind ourselves of the difference between a function and a functional. A function $\phi(x)$ is a mapping of a number x to a number $\phi(x)$. A functional $E[\phi]$ is a mapping of a function ϕ to the number $E[\phi]$. That is, the functional depends on the functional form of its argument. The functional derivative is denoted by

$$\frac{\delta E[\phi]}{\delta \phi(x)} \tag{19.16}$$

and is defined as follows. We make the small change

$$\phi \to \phi + \delta\phi_x \tag{19.17}$$

where $\delta\phi_x$ is a function that is almost zero everywhere except in the neighborhood of the point x, where it has a small "pimple," as illustrated in Figure 19.3. The functional should have a corresponding small change $\delta E[\phi]$ proportional to $\delta\phi(x)$, and the coefficient is the functional derivative.

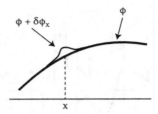

Figure 19.3 The function ϕ is varied in the neighborhood of x, by raising a small "pimple" at that point.

Some examples will show us how to carry out functional differentiations:

$$\frac{\delta}{\delta\phi(x)} \int dy\, f(y)\phi^n(y) = n\phi^{n-1}(x)f(x)$$

$$\frac{\delta}{\delta\phi(x)} \int dy\, f(y)\frac{\partial}{\partial y}\phi(y) = -\frac{\partial}{\partial x}f(x) \qquad (19.18)$$

In the last equation, we have assumed that $f(y) = 0$ at the boundaries of the y-integration, or that periodic boundary conditions are imposed.

19.5 Second-Order Phase Transition

In a second-order phase transition, as exemplified by spontaneous magnetization, the system suddenly magnetizes below a critical temperature T_c in the absence of an external field, as illustrated in Figure 19.4. In the presence of an external field, however, induced magnetization is always present, and there is no sharp transition.

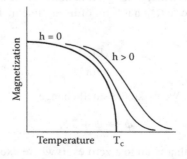

Figure 19.4 Spontaneous magnetization of a ferromagnet. The sharp phase transition happens only in the absence of external field. Otherwise an induced magnetization is always present.

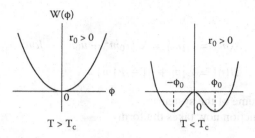

Figure 19.5 The potential that models a second-order phase transition. As the temperature decreases below the critical temperature, it develops two minima at nonzero values of the order parameter $\pm\phi_0$. The system chooses only one of these, in a phenomenon known as "spontaneous symmetry breaking."

Such behavior can be modeled by choosing the potential to have the form

$$W(\phi) = r_0\phi^2 + u_0\phi^4 \tag{19.19}$$

This is invariant under a sign change in ϕ, reflecting the intrinsic rotational invariance of a ferromagnet. As illustrated in Figure 19.5, this function has only one minimum when $r_0 > 0$, but develops two minima when $r_0 < 0$. We utilize this feature, by making r_0 change sign at the critical temperature:

$$r_0 = bt \tag{19.20}$$

where b is a real positive constant, and t is the fractional deviation from the critical temperature T_c:

$$t = \frac{T - T_c}{T_c} \tag{19.21}$$

The parameter u_0 is assumed to be independent of temperature, and must be positive to maintain stability.

19.6 Mean-Field Theory

In a uniform external field h, the mean field is independent of x:

$$\phi_0(x) = m \tag{19.22}$$

which is the uniform magnetization density that minimizes $E[\phi]$. If we write a general field in the form $\phi(x) = m + \eta(x)$, then

$$E[\phi] = A(m) + B[\eta] \tag{19.23}$$

with

$$A(m) = E[m] = V\left(r_0 m^2 + u_0 m^4 - hm\right)$$

$$B[\eta] = E[m + \eta] - E[m] \tag{19.24}$$

where V is the volume of the system.

The partition function now takes the form

$$Q = e^{-\beta A(m)} \int D\eta e^{-\beta B[\eta]} \tag{19.25}$$

and thus

$$\text{Free energy} = A(m) - \frac{1}{\beta} \ln \int D\eta e^{-\beta B[\eta]} \tag{19.26}$$

The first term is the contribution from the mean field, and the second term arises from fluctuations about the mean field. Neglecting the latter leads to "mean-field theory," in which the free energy is simply $A(m)$. This is a convenient starting point for investigations, because it is very simple, and because it is able to delineate the possible phases of the system.

The value of the mean field m is determined by minimizing $A(m)$, leading to the equation

$$2r_0 m + 4u_0 m^3 - h = 0 \tag{19.27}$$

For $h = 0$ we have

$$m\left(m^2 + \frac{r_0}{2u_0}\right) = 0 \tag{19.28}$$

with the possible roots $0, \pm\sqrt{-r_0/2u_0}$. The roots $\pm\sqrt{-r_0/2u_0}$ are acceptable only if $r_0 < 0$, because m must be real. We can verify that for $r_0 < 0$ the root $m = 0$ corresponds to a maximum instead of a minimum. Thus we have

$$m = \begin{cases} 0 & (r_0 > 0) \\ \pm\sqrt{-r_0/2u_0} & (r_0 < 0) \end{cases} \tag{19.29}$$

For $r_0 < 0$, we must choose between the two minima, and let us pick the positive root. (We can tilt the system in favor of this positive root by keeping an infinitesimally small positive external field.) Then, with Equation (19.20), we have

$$m = \begin{cases} 0 & (t > 0) \\ \sqrt{(b/2u_0)t} & (t < 0) \end{cases} \tag{19.30}$$

where $t = (T - T_c)/T_c$. The phase transition is second order, in that m is continuous at the critical point $t = 0$, but its slope is discontinuous.

We have thus described a phase transition involving a discontinuous change, in terms of continuous functions.[2] This comes about because of the bifurcation of the minimum of the potential, when the temperature is varied. The fact that the system must choose one of the two possible minima is an example of "spontaneous symmetry breaking," which we will discuss in more detail in Section 20.2.

19.7 Critical Exponents

At the critical point, a thermodynamic function generally contains a term regular in t, plus a singular part that behaves like a power of t. The power is called the *critical exponent*. The following exponents α, β, γ are define at $h = 0$ as $t \to 0$:

$$M \sim |t|^{\beta} \quad \text{(order parameter)}$$

$$\chi \sim |t|^{-\gamma} \quad \text{(susceptibility)}$$

$$C \sim |t|^{-\alpha} \quad \text{(heat capacity)} \tag{19.31}$$

where $\sim$ means "singular part is proportional to." (Note that β here is not $(k_B T)^{-1}$.) These exponents should be the same whether we approach the critical point from above or below; but the proportional constant can be different, and may be zero on one side. Another exponent δ is defined at $t = 0$ as $h \to 0^+$:

$$M \sim h^{1/\delta} \quad \text{(equation of state)} \tag{19.32}$$

Critical exponents are interesting because they have universality, being shared by a class of systems. We shall expand on this point in the next sections.

For now, let us calculate the critical exponents in the mean-field theory. From Equation (19.30) we immediately obtain

$$\beta = \frac{1}{2} \tag{19.33}$$

At $t = 0$ and $h > 0$, we substitute the value $r = 0$ into Equation (19.27), and obtain $4u_0 m^3 - h = 0$. Therefore

$$\delta = 3 \tag{19.34}$$

To calculate the susceptibility χ, we differentiate both sides of Equation (19.27) with respect to h, obtaining

$$\chi = \frac{1}{2bt + 12u_0 m^2} \tag{19.35}$$

---

[2]This idea is the basis of the so-called "catastrophe theory." For introduction, see Arnold (1984).

Substituting Equation (19.30) into the right side, we have

$$
\chi = \begin{cases} (2bt)^{-1} & (t > 0) \\ (6bt)^{-1} & (t < 0) \end{cases}
\tag{19.36}
$$

Hence

$$
\gamma = 1
\tag{19.37}
$$

For $h = 0$ and $t \to 0$, we have from Equations (19.24) and (19.30)

$$
\frac{A}{V} = \begin{cases} 0 & (t > 0) \\ (3b^2/u_0)t^2 & (t < 0) \end{cases}
\tag{19.38}
$$

from which we obtain, using Equation (19.10),

$$
\frac{C}{V} = \begin{cases} 0 & (t > 0) \\ -6b^2/(u_0 T_c^2) & (t < 0) \end{cases}
\tag{19.39}
$$

This gives

$$
\alpha = 0
\tag{19.40}
$$

19.8   The Correlation Length

Going beyond mean-field theory, we must consider spatial variations of the order parameter. Uniformity is the equilibrium situation, since it is favored by the kinetic term in the Landau free energy. If we perturbed the uniformity by introducing a point-like disturbance at any point, we expect that the density will return to uniformity as we go away from that point. There is a characteristic "healing" distance, which is called the *correlation length* and is a property of the *correlation function* defined by

$$
G(x, y) = \langle \phi(x)\phi(y) \rangle - \langle \phi(x) \rangle \langle \phi(y) \rangle
\tag{19.41}
$$

If there is no correlation between the values of the field at x and y, we would have $G(x, y) = 0$, because the joint average $\langle \phi(x)\phi(y) \rangle$ would be the equal to the product of the individual averages. For large separations $|x - y| \to \infty$, we expect the behavior

$$
G(x, y) \sim \exp\left(-\frac{|x - y|}{\xi}\right)
\tag{19.42}
$$

This defines the correlation length ξ, which will depend on the temperature.

We can estimate the correlation length in the mean-field approximation, by minimizing the Landau free energy in the presence of an external potential concentrated at the origin, of the form

$$
h(x) = h_0 \delta^D(x)
\tag{19.43}
$$

The Landau free energy is then

$$E[\phi] = \int d^D x \left[\frac{1}{2}|\nabla\phi(x)|^2 + r_0\phi^2(x) + u_0\phi^4(x) - h_0\phi(x)\delta^D(x)\right] \quad (19.44)$$

The term $|\nabla\phi(x)|^2$ may be replaced by $-\frac{1}{2}\phi(x)\nabla^2\phi(x)$, by performing a partial integration. The mean field $m(x)$ is now nonuniform, and serves as an estimate of $G(x, 0)$.

Consider a small variation about the mean field:

$$\phi(x) = m(x) + \delta\phi(x) \quad (19.45)$$

Since $m(x)$ is supposed to minimize $E[\phi]$, the variation in the latter should be of second-order smallness in $\delta\phi$. Thus,

$$0 = \delta E[\phi] = \int d^D x \left[-\nabla^2 m(x) + 2r_0 m^2(x) + 4u_0 m^3(x) - h_0\delta^D(x)\right]\delta\phi(x) \quad (19.46)$$

Since $\delta\phi$ is arbitrary, we obtain the following differential equation:

$$-\nabla^2 m(x) + 2r_0 m(x) + 4u_0 m^3(x) = h_0\delta^D(x) \quad (19.47)$$

This equation occurs in such diverse fields as plasma physics, quantum optics, superfluidity, and the theory of elementary particles, and is called the inhomogeneous "nonlinear Schrödinger equation" (NLSE).

To get a solvable equation, we drop the nonlinear m^3 term, arguing that m is small for $t > 0$. The equation becomes

$$-\nabla^2 m(x) + 2r_0 m(x) = h_0\delta^D(x) \quad (19.48)$$

Taking the Fourier transform of both sides we obtain

$$(k^2 + 2r_0)\tilde{m}(k) = h_0 \quad (19.49)$$

where $\tilde{m}(k)$ is the Fourier transform of $m(x)$:

$$\tilde{m}(k) = \int d^D x e^{-ik\cdot x} m(x)$$

$$m(x) = \int \frac{d^D k}{(2\pi)^D} e^{ik\cdot x} \tilde{m}(k) \quad (19.50)$$

Thus

$$\tilde{m}(k) = \frac{h_0}{k^2 + 2r_0} \quad (19.51)$$

and the inverse transform gives

$$m(x) = h_0 \int \frac{d^D k}{(2\pi)^D} \frac{e^{ik\cdot x}}{k^2 + 2r_0} \quad (19.52)$$

For for $D > 2$ the asymptotic behavior for large $|x|$ is

$$m(x) \approx C_0 |x|^{2-D} \exp\left(-\frac{|x|}{\xi}\right) \tag{19.53}$$

where C_0 is a constant, and the correlation length ξ is given by

$$\xi = (2r_0)^{-1/2} = (2bt)^{-1/2} \tag{19.54}$$

Its behavior at the critical point defines the critical exponent ν:

$$\xi \sim t^{-\nu} \tag{19.55}$$

According to our calculation

$$\nu = \frac{1}{2} \tag{19.56}$$

This is a value in the mean-field approximation, since we have ignored fluctuations about $m(x)$.

At the critical temperature ξ diverges, and the exponential law is replaced by a power law. This means that the exponential law governing the decay of inhomogeneity is replaced by a power law $|x|^{2-D}$, according to Equation (19.53). This is a mean-field approximation, however. More accurate analyses gives

$$m(x) \sim |x|^{2-D-\eta} \quad (t = 0) \tag{19.57}$$

The dimension of space seems to have changed from D to $D + \eta$, which is a critical exponent called the "anomalous dimension." This case $D = 2$ is very interesting, leading up to vortex-antivortex creation and annihilation; but that is beyond the scope of our discussion (Huang 1998).

The physical import of the correlation length is that the system organizes itself into more or less uniform blocks of size ξ. Thus, we cannot resolve spatial structures on a finer scale than ξ. As we approach the critical point, ξ increases, and we lose resolution. At the critical point, when ξ diverges, we cannot see any details at all. Only global properties, such as the dimension of space, or the number of degrees of freedom, distinguish one system from another. That is why systems at the critical point fall into universality classes characterized by a shared set of critical exponents.

Imagine that you are blind-folded in a room, which you can probe only with a very long pole. You can find out whether the room is 1D, 2D, or 3D by trying to move the pole, but you can learn little else. Similarly, if you put on very dark eyeglasses, you would conclude that all places on Earth are the same, characterized by a 24-hour light-dark cycle. To experience something new, you would have to go to Mars. (Or take those glasses off.)

19.9 First-Order Phase Transition

We can describe a first-order transition in the context of Ginsburg–Landau theory as follows. We regard ϕ as the spatial density of the system, and construct a potential W such that, as the temperatures varies, the function assumes a sequence of shapes shown qualitatively in Figure 19.6. The two minima at ϕ_1, ϕ_2 represent respectively the low-density and high-density phase of the system. Just below the transition temperature, ϕ_1 corresponds to the stable phase. At the transition point, both ϕ_1 and ϕ_2 are stable, and just above the transition point ϕ_2 becomes the stable phase.

We can choose W is any convenient way to suit our purpose, but it is interesting to note that W can be so chosen that we obtain the van der Waals equation of state. For a uniform ϕ, in the absence of external field, the Landau free energy takes the form

$$E(\phi) = VW(\phi) \tag{19.58}$$

where the total volume V is a fixed parameter. Thus, the potential $W(\phi)$ is the free energy per unit volume. The pressure is given by the Maxwell relation

$$P = -\frac{\partial W(\phi)}{\partial(1/\phi)} = \phi^2 W'(\phi) \tag{19.59}$$

where $W'(\phi) = \partial W/\partial\phi$. Equating this to the van der Waals equation of state leads to

$$W'(\phi) = \frac{RT}{\phi(1 - b\phi)} - a$$

$$W(\phi) = RT \ln\frac{\phi}{1 - b\phi} - a\phi + c \tag{19.60}$$

where c is a constant.

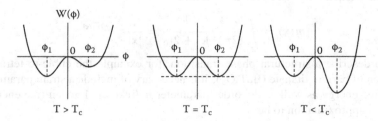

Figure 19.6 Modeling a first-order phase transition.

19.10   Cahn–Hilliard Equation

In the context of the Ginsburg–Landau theory, the relaxation of spatial inhomogeneity can be treated as follows. The local free energy density can be obtained from the Landau free energy by functional differentiation:

$$\frac{\delta E[\phi]}{\delta \phi(x)} = -\varepsilon^2 \nabla^2 \phi(x) + W'(\phi(x)) \tag{19.61}$$

where $W'(\phi) = \partial W(\phi)/\partial \phi$. We assume that a spatial gradient in this quantity will induce a diffusion current proportional to the gradient:

$$\mathbf{j}(x) = -\nabla[-\varepsilon^2 \nabla^2 \phi + W'(\phi)] \tag{19.62}$$

The proportionality constant has been absorbed into ε and W. If there is no dissipation, the current density must satisfy the continuity equation

$$\frac{\partial \phi(x)}{\partial t} + \nabla \cdot \mathbf{j} = 0 \tag{19.63}$$

Combining this with the previous equation, we obtain the Cahn–Hilliard equation

$$\frac{\partial \phi(x)}{\partial t} = -\nabla^2[\varepsilon^2 \nabla^2 \phi - W'(\phi)] \tag{19.64}$$

Dissipation will add an extra term to the right side, but must come from assumptions beyond the Ginsburg–Landau theory.

A interesting application is the numerical simulation of spinodal decomposition. One chooses $W(\phi)$ appropriate for a first-order transition, as in Figure 19.6, and numerically solves the equation with an initial configuration at an unstable point, such as near $\phi = 0$ in Figure 19.6. Results of such simulations (Zhu and Chen 1999) yield the sequence of pictures in Figure 4.12.

Problems

19.1 Verify Equation (19.53) for $D = 3$, that is

$$m(\mathbf{x}) = \int \frac{d^3 k}{(2\pi)^3} \frac{e^{i\mathbf{k} \cdot \mathbf{x}}}{\mathbf{k}^2 + 2r_0} = \frac{1}{|\mathbf{x}|} e^{-\sqrt{2r_0}|\mathbf{x}|}$$

19.2 To describe a structural phase transition, for example the cubic-to-tetragonal transition of barium titanate ($BaTiO_3$), it is necessary of include a strain parameter ε in the free energy as well as the order parameter η. Take the Landau free energy in the mean approximation to be

$$E(\eta, \varepsilon) = at\eta^2 + b\eta^4 + c\varepsilon^2 + g\eta^2\varepsilon$$

where $t = T - T_c$, and a, b, c, g are positive constants. In particular, g is called the coupling constant.

(a) Minimize the free energy with respect to ε for fixed η, to determine $\bar{\varepsilon}$ as a function of η.

(b) Obtain the effective free energy for η alone. What is the new "renormalized" value $\tilde{b}$ of b?

(c) What happens to $\tilde{b}$ as a function of g? What happens to the phase transition as a function of g?

19.3 The nematic liquid crystal used in displays can be described by an order parameter S corresponding to the degree of alignment of molecular directions. In the ordinary fluid phase $S = 0$. The transition between the ordinary fluid phase and a nematic phase can be modeled via the mean-field Landau free energy

$$E(S) = at S^2 + bS^3 + cS^\$$$

where $t = T - T_0$, and a, b, c are positive constants. The third-order coefficient b is usually small.

(a) Sketch $E(S)$ for range of T, from $T \gg T_0$, through $T = T_0$ to $T < T_0$. Comment on the value of S at the minimum of $E(S)$ at each value of T considered.

(b) What are the conditions for $E(S)$ to be minimum?

(c) Find the transition temperature T_c. (It is not T_0.)

(d) Is there a latent heat associated with the phase transition? If so, what is it?

(e) How does the order parameter vary below T_c?

19.4 A system has a two-component order parameter: $\{\phi_1, \phi_2\}$. The mean-field Landau free energy is

$$E(\phi_1, \phi_2) = E_0 + at\left(\phi_1^2 + \phi_2^2\right) + b\left(\phi_1^2 + \phi_2^2\right)^2 + c\left(\phi_1^4 + \phi_2^4\right)$$

where $t = T - T_c$, and a, b are positive constants.

(a) Represent the order parameter as a vector on a plane. Use polar coordinate to write $\phi_1 = \phi \cos\theta$, $\phi_1 = \phi \sin\theta$, where $\phi = \sqrt{\phi_1^2 + \phi_2^2}$. Minimize the free energy with respect to ϕ, and show that its minimum occurs at the minimum of

$$\tilde{b} = b + c(\cos^4\theta + \sin^4\theta)$$

(b) Find the values of the order parameter in the ordered phase for the three possibilities $c < 0, c = 0, c > 0$.

(c) For $c < 0$, suppose the order parameter is parallel to the x axis. Calculate the susceptibility χ_x with respect to an external field applied in the x direction, and χ_y in the y direction.

19.5 Broken symmetry A highway is proposed to connect four cities located at the corners of a unit square, as shown in the upper panel of the accompanying sketch. Show that the total length of the system can be reduced by adopting either of the schemes in the lower panel. Find the distance a that minimizes the length. The invariance of the square under a 90° rotation is expressed through the fact that there are two minimal schemes that go into each other under the rotation.

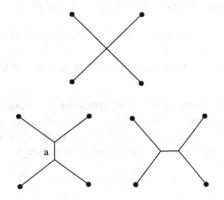

References

Huang, K., *Quantum Field Theory, From Operators to Path Integrals*, Wiley, New York, 1998, Chapter 18.

van der Waals, J.D. "The thermodynamic theory of capillarity flow under the hypothesis of continuous variation in density," *Verhandelingen der Koninklijke Nederlansche Akademie van Wetenchappen te Amsterdam*, **1**:1–56 (1893).

Zhu, J. and L.-Q. Chen, *Phys. Rev.*, **E60**:3564 (1999).

Chapter 20

Superfluidity

20.1 Condensate Wave Function

In the ideal Bose gas, a finite fraction of the particles occupies the same single-particle state below a critical temperature, forming a Bose–Einstein condensate. When the particles have mutual interactions, however, single-particle states are no longer meaningful. Nevertheless, we can still define a *condensate wave function* $\psi(\mathbf{r})$, as the quantum amplitude for removing a particle from the condensate at position $\mathbf{r}$. Its complex conjugate $\psi^*(\mathbf{r})$ is the amplitude for creating a particle.

We can imagine creating a particle in the condensate by inducing a transition into it. As we saw in Section 16.2, Bose enhancement makes the rate proportional to the existing boson density. Thus, the creation amplitude $\psi^*(\mathbf{r})$ should be proportional to the square root of the density. We define $|\psi(\mathbf{r}, t)|^2 d^3r$ as the number of condensate particles in the volume element d^3r, so that the total number of particles in the condensate is

$$\int d^3r \, |\psi(\mathbf{r})|^2 = N_0 \tag{20.1}$$

In Chapter 18, we were able to give a complete mathematical description of the Bose–Einstein condensation in the ideal gas. Such a treatment is not possible when there is interaction between particles. Here, we shall describe it in Ginsburg–Landau theory, with the complex field ψ as order parameter. The Landau free energy is taken to be

$$E[\psi, \psi^*] = \int d^3r \left[\frac{\hbar^2}{2m} |\nabla\psi|^2 + (U_{\text{ext}} - \mu)\psi^*\psi + \frac{g}{2}(\psi^*\psi)^2 \right] \tag{20.2}$$

where $|\nabla\psi|^2 = \nabla\psi^* \cdot \nabla\psi$, m is the mass of the particles, $U_{\text{ext}}(\mathbf{r})$ is an external potential, and μ is the chemical potential that determines N_0. There are equivalent choices for independent variables:

- ψ and its complex conjugate ψ^*,
- real and imaginary parts Re ψ and Im ψ,
- modulus and phase angle $\sqrt{n_0}$ and φ in the polar representation

$$\psi(x) = \sqrt{n_0(x)} \, e^{i\varphi(x)}$$

where n_0 is the density of the condensate.

In the Ginsburg–Landau philosophy, Equation (20.2) is to be used near the critical temperature T_c, where ψ is small. On the other hand, the form of Equation (20.2) can be derived from a microscopic theory as the effective Hamiltonian for a dilute Bose gas at low temperatures (Huang 1987), and one can relate the parameter g to atomic properties:

$$g = \frac{4\pi a \hbar^2}{m} \qquad (20.3)$$

where a is the "scattering length," an equivalent hard-sphere diameter of an atom. The conditions for the validity of this picture are

$$n_0^{1/3} a \ll 1$$

$$\frac{a}{\lambda} \ll 1 \qquad (20.4)$$

where $n_0 = N_0/V$, and $\lambda = \sqrt{2\pi\hbar^2/mk_BT}$ is the thermal wavelength. Thus, we can use Equation (20.2) in two separate neighborhoods: near $T = T_c$ or $T = 0$.

The grand partition function of the condensate is given by

$$\mathcal{Q} = \int D\psi\, D\psi^* e^{-\beta E[\psi,\psi^*]} \qquad (20.5)$$

where $\int D\psi D\psi^* = \int D(\mathrm{Re}\psi)D(\mathrm{Im}\psi)$ denotes functional integrations over the real and imaginary parts of ψ independently. The mean-field theory amounts to assuming that one configuration dominates the integral, namely the "ground state" that minimizes the free energy $E[\psi, \psi^*]$ and ignores all fluctuations.

20.2 Spontaneous Symmetry Breaking

The potential term in Equation (20.2) has the form

$$W(\psi\psi^*) = \frac{g}{2}|\psi|^4 - \mu|\psi|^2 \qquad (20.6)$$

which depends only on $|\psi|^2 = \psi^*\psi$. In Figure 20.1 we show a plot of this over the complex ψ-plane, for different signs of μ. If $\mu < 0$. the minimum of the potential occurs at $\psi = 0$. For $\mu > 0$, however, the potential has a wine-bottle shape, with a continuous distribution of minima along a trough, along the dashed circle in Figure 20.1. We can model the Bose–Einstein condensation as a second-order phase transition by making μ dependent on temperature, and change sign at the critical temperature.

The Landau free energy is invariant under the phase transformation

$$\psi \to e^{i\chi}\psi$$

$$\psi^* \to e^{-i\chi}\psi^* \qquad (20.7)$$

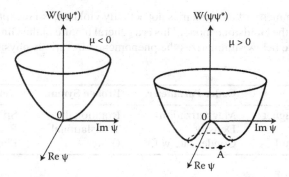

Figure 20.1 If $\mu > 0$, the ground state corresponds to a definite location A on the dotted circle, which is the same over all space (when there is no external potential). This state has a definite phase, and breaks global gauge invariance. The Goldstone mode is the excited state in which A runs along the dotted circle as the spatial location changes.

where χ is a constant. This symmetry is known as *global gauge invariance*, and leads to the conservation of particles. When $\mu > 0$, the ground state wave function has the form $\sqrt{n_0}\, e^{i\varphi}$ over all space (for the case $U_{\text{ext}} = 0$). This corresponds to the same point A on the dotted circle in Figure 20.1, over all space. This breaks the global gauge invariance, and is called *spontaneous symmetry breaking*, because it is not induced by any external agent.

Accompanying the occurrence of spontaneous symmetry breaking is a special mode of excitation called the *Goldstone mode*. In Figure 20.1, it corresponds to choosing A at different points around the trough, as the spatial position changes. Figure 20.2 illustrates this mode in a spin analogy. The orientation of a spin represents the phase φ at a particular spatial location. In the ground state, the spins are parallel to each other, but the orientation is arbitrary. In the Goldstone mode, the spin orientations change as the position changes. The excitation energy of this mode is due to kinetic energy, and vanishes as the wavelength of variation approaches infinity.

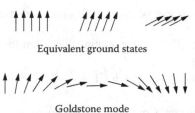

Equivalent ground states

Goldstone mode

Figure 20.2 Spin analog of the Goldstone mode. The orientiation of a spin represents the phase angle at a particular location. In the ground state they all point in the same direction, which is however arbitrary. The Goldstone mode is the excited state in which the spin turns as the location changes.

The basic symmetry of the system is not actually violated, but reexpressed through the existence of the Goldstone mode. This is a general feature of the Ginsburg–Landau theory. The table below summarizes the phenomenon in various physical systems:

System	Order Parameter	Broken Symm.	Goldstone
Ferromagnet	Magnetization	Rotational	Spin wave
Solid	Density	Translational	Phonon
Superfluid	Condensate w.f.	Gauge	Phonon

20.3 Mean-Field Theory

In the mean-field approximation, we take $\langle \psi \rangle = \psi_0$, which is the uniform wave function that minimizes the Landau free energy. For simplify we will just write ψ, and drop the subscript 0. In the absence of an external potential

$$E[\psi, \psi^*] = V \left[\frac{g}{2} |\psi|^4 - \mu |\psi|^2 \right] \tag{20.8}$$

Minimization should be carried out with respect to Re ψ and Im ψ independent, and it leads to

$$|\psi| = \begin{cases} 0 & (\mu < 0) \\ \sqrt{\mu/g} & (\mu > 0) \end{cases} \tag{20.9}$$

Thus, $\mu < 0$ corresponds to the high-temperature phase where there is no condensate. When $\mu > 0$, a condensate forms. The normalization condition (Equation [20.1]) gives

$$\mu = \frac{g N_0}{V} \tag{20.10}$$

In contrast to the ideal Bose gas, where $\mu = 0$ in the condensed phase, now it depends on the condensate density.

In the neighborhood of $T = T_c$, we assume that μ has the form

$$\mu = b \left(1 - \frac{T}{T_c} \right) \tag{20.11}$$

where b is a positive constant. Comparison with Equation (20.10) yields

$$\frac{N_0}{V} = \begin{cases} 0 & (T > T_c) \\ \dfrac{b}{g} \dfrac{T}{T_c} & (T < T_c) \end{cases} \tag{20.12}$$

In this approximation, all the critical exponents have the mean-field given in Section 19.7.

When there is an external potential, ψ must vary in space. Minimization of the Landau free energy (Equation [20.2]) leads to a nonlinear Schrödinger equation (NLSE):

$$\left[-\frac{\hbar^2}{2m}\nabla^2 + U + g|\psi|^2\right]\psi = \mu\psi \tag{20.13}$$

which in the present context is known as the *Gross–Pitaevsky equation*. It can be generalized to the time-dependent form

$$\left[-\frac{\hbar^2}{2m}\nabla^2 + U + g|\psi|^2\right]\psi = i\hbar\frac{\partial\psi}{\partial t} \tag{20.14}$$

which describes the Goldstone mode, as we see later. The normalization $\int d^3r|\psi|^2$ is a constant of the motion. It is fixed by g, and cannot be chosen arbitrarily.

20.4 Observation of Bose–Einstein Condensation

Bose–Einstein condensation has been experimentally achieved in dilute gases of bosonic alkali atoms confined in an external potential (Ketterle et al. 1999). The condensate typically contains the order of 10^6 particles, with a spatial extension of order 10^{-2} cm. The average density is of order 10^{12} cm^{-3}, and the transition temperature is of order 10^{-6} K. Compared with liquid ^{4}He, the density is smaller by ten orders of magnitude, and the transition temperature lower by seven orders of magnitude.

Figure 20.3 shows the observed density profile in a gas of Na atoms trapped in a harmonic potential, for a series of temperatures. At the t transition temperature 1.7×10^{-6} K, a central peak begins to form, representing the squared modulus $|\psi|^2$ of the condensate wave function. It continues to grow with decreasing temperature, acquiring atoms from the surrounding thermal cloud. Below 0.5×10^{-6} K almost all atoms are in the condensate.

We can neglect the kinetic term in Equation (20.13), where the wave function varies slowly in space, and obtain the approximate formula

$$|\psi(\mathbf{r})|^2 \approx \frac{1}{g}[\mu - U_{\text{ext}}(\mathbf{r})] \tag{20.15}$$

This is called the "Thomas–Fermi approximation" after an approximation with that name in atomic physics. In the data of Figure 20.3 the condensate has an extension of about 0.03 cm. This is to be compared with 0.005 cm, the extent of the free-particle wave function in the potential. We see that the repulsion between atoms considerably broadens the condensate peak.

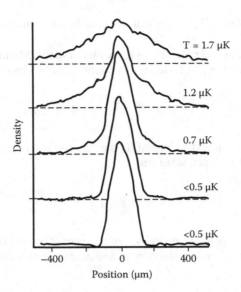

Figure 20.3   Observation of Bose–Einstein condensation in a trapped atomic Na gas. The density profile is measured at successively lower temperatures. Bose–Einstein condensation occurs at 1.7 μK, when a central peak emerges, representing the squared modulus of the condensate wave function. (After Stenger et al., 1998.)

20.5   Quantum Phase Coherence

The order parameter embodies the idea of the macroscopic occupation of a single mode, and this implies quantum phase coherence. That is to say, if you imagine pulling particles out of the condensate one by one, you would find that their wave functions all have the same quantum phase.

Consider two condensates moving toward each other, that eventually overlap, as illustrated in Figure 20.4. Each has a wave function that is approximately a plane

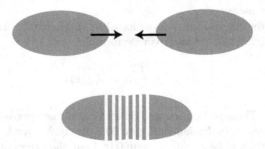

Figure 20.4   Two condensates approach each other, overlap, and exhibit interference fringes.

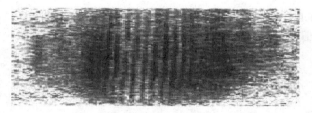

Figure 20.5 Photograph of interference fringes produced by two overlapping condensates of Na atoms. The separation between fringes is 1.5×10^{-3} cm. (Courtesy of W. Ketterle.)

wave, except near the edges of the condensate. The total wave function is the sum of the two wave functions, since an atom can belong to either condensate:

$$\psi(r, t) = C_1 e^{i(\mathbf{k}_1 \cdot \mathbf{r} - \omega_1 t)} + C_2 e^{i(\mathbf{k}_2 \cdot \mathbf{r} - \omega_2 t)} \tag{20.16}$$

where the subscript $i = 1, 2$ labels the two condensates, $\hbar \omega_i = \hbar^2 k_i^2 / 2M_i$, where M_i is the mass of the condensate, and C_i is a constant. The density of the total system is given by

$$|\psi(r, t)|^2 = |C_1|^2 + |C_2|^2 + 2\mathrm{Re}C_1^* C_2 e^{i(\mathbf{k}_2 - \mathbf{k}_1) \cdot \mathbf{r} - i(\omega_2 - \omega_1)t} \tag{20.17}$$

The last term exhibits interference fringes, which are visible in the photograph in Figure 20.5 produced by interference between two Na condensates.

The condensate in the ideal Bose gas does not exhibit phase coherence. The condensate wave function in this case is a product of single-particle wave functions with no correlations among them:

$$\Psi_0 = u_0(\mathbf{r}_1) \cdots u_0(\mathbf{r}_N) \tag{20.18}$$

where the phase of each wave function u_0 can be chosen arbitrarily, for the choices merely affect the overall phase factor, which is arbitrary. When there are interactions, no matter how weak, then there is a correction to the free-particle wave function:

$$\Psi_0 = [u_0(\mathbf{r}_1) \cdots u_0(\mathbf{r}_N)] + \Psi' \tag{20.19}$$

The relative phase between the first term and the second is now determined by the interactions, and we cannot change it arbitrarily. Thus, phase coherence is a result of interactions, just as spontaneous magnetization is a result of spin-spin attractions.

20.6 Superfluid Flow

The time-dependent NLSE (Equation [20.14]) conserves the number of particles in the condensate, and we have the continuity equation

$$\frac{\partial n}{\partial t} + \nabla \cdot \mathbf{j} = 0 \tag{20.20}$$

where n is the particle density, and $\mathbf{j}$ the particle current density:

$$\mathbf{j}(\mathbf{r}) = \frac{\hbar}{2mi}(\psi^*\nabla\psi - \psi\nabla\psi) \qquad (20.21)$$

Putting $\psi = \sqrt{n_0}e^{i\varphi}$, we have

$$\mathbf{j} = \frac{n\hbar}{m}\nabla\varphi \qquad (20.22)$$

from which we identify the *superfluid velocity*

$$\mathbf{v}_s = \frac{\hbar}{m}\nabla\varphi \qquad (20.23)$$

which describes a flow of the condensate without dissipation.

The circulation of the superfluid velocity field around a close path C is

$$\oint_C ds \cdot \mathbf{v}_s = \frac{\hbar}{m}\oint_C ds \cdot \nabla\varphi \qquad (20.24)$$

The integral on the right side is the change of the phase angle upon traversing the loop C, and must be an integer multiple of 2π, by continuity of the condensate wave function. Therefore the circulation is quantized:

$$\oint_C ds \cdot \mathbf{v}_s = \frac{\hbar\kappa}{m} \quad (\kappa = 0, \pm 1, \pm 2, \dots) \qquad (20.25)$$

A vortex is a flow pattern with a nonvanishing vorticity concentrated along a directed line called the vortex core. In Figure 20.6 we show a vortex line, in which the vorticity is contained in a linear core, and a vortex ring, whose core is a closed curve. The circulation is zero around any closed loop not enclosing the vortex core, and nonzero otherwise, as illustrated by the loop C in the figure.

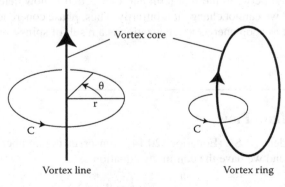

Figure 20.6   Vortex line and vortex ring.

For the vortex line, cylinder symmetry dictates that the velocity field $v_s(r)$ depends only on the normal distance r from the vortex core. Using Equation (20.25), we have

$$\int_0^{2\pi} d\theta\, r v_s(r) = \frac{\hbar\kappa}{m}$$ (20.26)

Thus

$$v_s(r) = \frac{\hbar\kappa}{mr}$$ (20.27)

which, according to Equation (20.23), gives

$$\varphi = \kappa\theta$$ (20.28)

where θ is the angle around the vortex core. This phase is defined only as $\mathrm{mod}(2\pi)$, but the superfluid velocity is unique. When $r \to 0$, the continuum picture breaks down at some atomic distance, which provides a cutoff to Equation (20.27), and gives a finite radius to the vortex core.

20.7 Phonons: Goldstone Mode

The Goldstone mode is an excited solution to the time-dependent NLSE (Equation [20.14]). We outline the steps to obtain it and quote results.

Take $U_{\text{ext}} = 0$. The ground state wave function is uniform in space, given by

$$\psi_0 = \sqrt{n_0}\exp(-i\mu t/\hbar)$$ (20.29)

where n_0 is the condensate density, and $\mu = gn$. For an excited state, we postulate the wave function

$$\psi(x, t) = \psi_0 + f(x, t)\exp(-i\mu t/\hbar)$$ (20.30)

Substituting this into Equation (20.14), and assuming that $f(x, t)$ is small, we obtain the following linear approximation to the equation:

$$i\hbar\frac{\partial f}{\partial t} = -\frac{\hbar^2}{2m}\frac{\partial^2 f}{\partial x^2} + gn(f + f^*)$$ (20.31)

which can be solved by putting

$$f(x, t) = Ue^{-i(\omega-kx)} + V^*e^{i(\omega-kx)}$$ (20.32)

where U and V^* are complex constants. The equation then imposes a relation between ω and k:

$$\omega = \frac{\hbar k}{2m}\sqrt{k^2 + 16\pi a n_0}$$ (20.33)

This is called the *Bogoliubov spectrum*, and represents the dispersion relation between energy $\hbar\omega$ and momentum $\hbar k$ of the Goldstone mode. In the long wavelength limit $k \to 0$, we have a phonon spectrum:

$$\omega = ck \tag{20.34}$$

where c is the sound velocity given by

$$c = \frac{\hbar}{2m}\sqrt{16\pi a n_0} \tag{20.35}$$

Problems

20.1 In the photograph in Figure 20.5, the interference fringes between two Bose–Einstein condensates have a spacing of 1.5×10^{-3}cm. Find the relative velocity between the two condensates.

20.2 Estimate the scattering length a of sodium atoms from the data contained in Figure 20.3, as follows. Use the Thomas–Fermi approximation (Equation [20.15]) for the condensate profile. Use Equation (20.10) for the chemical potential, with $N_0/V \approx 10^{11}$ cm^{-3}. The oscillator potential has a characteristic length $r_0 = \sqrt{\hbar/m\omega} = 5 \times 10^{-3}$ cm. The half width of the condensate is 3×10^{-2} cm.

20.3 Cold trapped atoms Consider a gas of N non-interacting nonrelativistic bosons of mass m in an external harmonic-oscillator potential in 3D. The Hamiltonian of a particle is

$$H = \frac{p^2}{2m} + \frac{1}{2}m\omega^2 r^2$$

where $p^2 = p_x^2 + p_y^2 + p_z^2$, and $r^2 = x^2 + y^2 + z^2$. Let $|\mathbf{n}\rangle$ be an eigenstate of H, where $\mathbf{n} = \{n_x, n_y, n_z\}$, with $n_x = 0, 1, 2, \ldots$, etc. The energy eigenvalues are

$$E_{\mathbf{n}} = \hbar\omega\left(n_x + n_y + n_z + \frac{3}{2}\right)$$

The fugacity z is determined through

$$N = \sum_{\mathbf{n}} \frac{1}{z^{-1}\exp(E_{\mathbf{n}}/k_B T) - 1}$$

The chemical potential is $\mu = k_B T \ln z$.
    (a) Show

$$\langle\mathbf{n}|r^2|\mathbf{n}\rangle = r_0^2\left(n_x + n_y + n_z + \frac{3}{2}\right)$$

where $r_0 = \sqrt{\hbar/m\omega}$.

Hint: Write x^2 in terms of creation and annihilation operators. (See Problem 10.1.)

(b) Prove the virial theorem

$$\frac{1}{2m} \langle \mathbf{n} | p^2 | \mathbf{n} \rangle = \frac{m\omega^2}{2} \langle \mathbf{n} | r^2 | \mathbf{n} \rangle = \frac{1}{2} E_{\mathbf{n}}$$

(c) Show that the thermal average of r^2 is

$$\langle r^2 \rangle = \frac{r_0^2}{N} \sum_{\mathbf{n}} \left(n_x + n_y + n_z + \frac{3}{2} \right) \frac{1}{z^{-1} \exp(E_{\mathbf{n}}/k_B T) - 1}$$

20.4 Continuing with the previous problem, estimate the transition temperature of the Bose–Einstein condensation, as follows. For $k_B T \gg \hbar\omega$, it is a good approximation to replace the sum over the quantum numbers by an integral.

Calculate the temperature T_0 at which $\mu = \frac{3}{2}\hbar\omega$. Show

$$k_B T_0 = b\omega N^{1/3}$$

where b is a numerical constant. This is not intensive because of the external potential. The $N^{1/3}$ dependence is verified by experiments, as shown in the accompanying figure. (After Mews et al., 1996.)

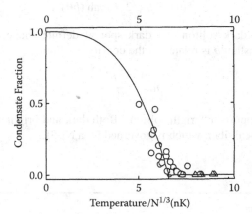

20.5 Continuing with the previous problem,

(a) Evaluate μ as a function of N and T in the classical limit, when the term -1 in the Bose distribution may be dropped. Show

$$\mu = \frac{3}{2}\hbar\omega + 3k_B T \ln \left(\frac{T_0}{bT} \right)$$

(b) Evaluate the mean-square radius $\langle r^2 \rangle$ of the trapped gas in the classical limit. Show

$$\langle r^2 \rangle = 3br_0^2 \left(\frac{T}{T_0} \right) N^{1/3}$$

(c) Make qualitative plots of μ and $\langle r^2 \rangle$ as functions of T, extrapolate down to T_0, and comment on what happens below T_0.

20.6 Soliton

Show that the 1D NLSE has a soliton solution.

Solution: From Equation (20.14)

$$i\hbar \frac{\partial \psi}{\partial t} = -\frac{\hbar^2}{2m} \frac{\partial^2 \psi}{\partial x^2} + g|\psi|^2 \psi$$

with $g = 4\pi a\hbar^2/m$, and $\int_{-\infty}^{\infty} dx |\psi|^2 = N$. Rewrite the equation in the form

$$i\frac{\partial \phi}{\partial \tau} = -\frac{\partial^2 \phi}{\partial x^2} + \lambda |\phi|^2 \phi$$

where $\phi = N^{1/2}\psi$, $\tau = (\hbar/2m)t$, and $\lambda = 8\pi aN$.

The following is a solution for $\lambda > 0$:

$$\phi(x,\tau) = f(\theta)e^{-i(\sqrt{2b}\theta + 4b\tau)}$$

where $\theta = x - v\tau$, $b = v^2/8$, and

$$f(\theta) = b\sqrt{2/\lambda}\,\tanh(b\theta)$$

This represents a "dark soliton"—a dark spot of defnite shape, propagating with velocity v. The constant b is related to the density at infinity:

$$n_\infty = \frac{b^2}{4\pi a}$$

The case $\lambda < 0$ supports a "bright soliton." Both dark and bright solitons have been observed in an optical fiber, which is governed by a NLSE.

References

Huang, K., *Statistical Mechanics*, 2nd ed., Wiley, New York, 1987, Section 13.8.

Ketterle, W., D.S. Durfee, and D.M. Stamper-Kurn, "Bose–Einstein condensation in atomic gases," *Proc. Int. School Phys.*, "Enrico Fermi," Course CXL, M. Inguscio, S. Stringari, and C.E. Wieman (eds.), IOS Press, Amsterdam, 1999, pp. 67–176.

Mews, M.-O. et al., *Phys. Rev. Lett.*, **77:**416 (1996).

Stenger, J. et al., *J. Low Temp. Phys.*, **113:**167 (1998).

Chapter 21

Superconductivity

21.1 Meissner Effect

Some metals make a second-order phase transition to a superconducting phase below a critical temperature. Electrical resistance seems to disappear in this phase, as the name suggests. The physical phenomenon arises from attractive interaction between electrons in the metal, induced by their coupling to lattice vibrations. The attractive force creates a bosonic bound state called the Cooper pair, and the transition is caused by the Bose–Einstein condensation of these bosons. One may view superconductivity as the superfluidity of the charged condensate. However, the Cooper pair is not a particle like an atom, but an extended object covering many lattice sites. Nevertheless, a superconductor can be described in Ginsburg–Landau theory.

The order parameter ψ of a superconductor is the complex wave function of the condensed Cooper pairs. We directly write down the NLSE, which is the generalization of Equation (20.14):

$$\left[-\frac{\hbar^2}{2m} \left(\nabla - \frac{2ie\hbar}{c} \mathbf{A} \right)^2 + a\psi + b|\psi|^2 \right] \psi = i\hbar \frac{\partial \psi}{\partial t} \qquad (21.1)$$

where a and b are constants. The system is coupled to a static vector potential $\mathbf{A}$, which corresponds to an external magnetic field $\mathbf{B} = \nabla \times \mathbf{A}$. The coupling constant e is the magnitude of the electron's charge. That is, the charge of the electron is $-e$, and the charge of a Cooper pair is $-2e$. This equation accounts for all the salient manifestations of superconductivity: the Meissner effect, magnetic flux quantization, and the Josephson effect.[1]

The equation is invariant under the gauge transformation

$$\mathbf{A} \to \mathbf{A} - \nabla \chi$$

$$\psi \to \psi e^{-i(2e/\hbar c)\chi} \qquad (21.2)$$

We impose the Coulomb gauge $\nabla \cdot \mathbf{A} = 0$. The current density is then given by

$$\mathbf{j} = \frac{e\hbar}{mi}(\psi^* \nabla \psi - \psi \nabla \psi) - \frac{4e^2}{mc}|\psi|^2 \mathbf{A} \qquad (21.3)$$

which satisfies $\nabla \cdot \mathbf{j} = 0$.

---

[1] For a more detailed discussion of the Ginsburg–Landau theory of superconductivity, see De Gennes 1966.

In a static uniform magnetic field, the order parameter is uniform, and the current density becomes

$$\mathbf{j} = -\frac{4e^2 n}{mc}\mathbf{A} \qquad (21.4)$$

where $n = |\psi|^2$ is the density of the condensate. This is know as the *London equation*. The vector potential satisfies Maxwell's equation

$$\nabla \times \nabla \times \mathbf{A} = \frac{4\pi}{c}\mathbf{j} \qquad (21.5)$$

Using $\nabla \times \nabla \times \mathbf{A} = \nabla(\nabla \cdot \mathbf{A}) - \nabla^2\mathbf{A}$ and $\nabla \cdot \mathbf{A} = 0$, we obtain

$$\left(\nabla^2 + \frac{16\pi e^2 n}{mc^2}\right)\mathbf{A} = 0 \qquad (21.6)$$

For uniform $\mathbf{A}$, we must have $\mathbf{A} = 0$. This means that the magnetic field must vanish inside a superconductor, and so does the current density. This is the *Meissner effect*.

If the superconductor is immersed in a uniform magnetic field, then the field can penetrate into the superconductor only within a layer of thickness called the *penetration depth*:

$$d = \sqrt{\frac{mc^2}{16\pi e^2 n}} \qquad (21.7)$$

A current density exists in this layer to shield the interior from the external magnetic field, and this makes up the superconducting current.

21.2   Magnetic Flux Quantum

Consider a hollow pipe made of a superconducting material like lead, with a total magnetic flux Φ inside the hollow, as shown in Figure 21.1. Inside the superconducting material the magnetic field must be zero. Thus, $\nabla \times \mathbf{A} = 0$, and the vector potential has "pure gauge" form

$$\mathbf{A} = \nabla\chi \qquad (21.8)$$

On the other hand, its line integral along the closed loop C must give the total flux:

$$\oint_C d\mathbf{s} \cdot \mathbf{A} = \Phi \qquad (21.9)$$

For a circular loop C, this can be satisfied by choosing

$$\chi = \frac{\theta\Phi}{2\pi} \qquad (21.10)$$

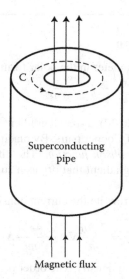

Figure 21.1 Magnetic flux inside the superconducting pipe is quantized in units of $\Phi_0 = \frac{hc}{e}$.

where θ is an angle around the loop. However, the vector potential must be removable through a gauge transformation, for otherwise there will be a current inside the superconductor according to Equation (21.4). Under such a gauge transformation, the order parameter undergoes the phase change

$$\psi \to \psi e^{-i(2e/\hbar c)\chi} = \psi e^{-i(2e\Phi/hc)\theta}$$

Since ψ must be periodic in θ with period 2, we must require

$$\frac{2e\Phi}{hc} = \kappa \tag{21.11}$$

where κ is an integer. This is the analog of vortex quantization in superfluids. The quantization condition can be rewritten as:

$$\Phi = \frac{\kappa}{2}\Phi_0 \tag{21.12}$$

where

$$\Phi_0 = \frac{hc}{e} \approx 10^{-7} \text{Gauss cm}^2 \tag{21.13}$$

is defined as the *magnetic flux quantum*.

The flux quantization is an equilibrium condition. If we create an arbitrary flux through the pipe in Figure 21.1, by suddenly thrusting the pipe into an arbitrary preexisting magnetic field, for example, supercurrent will be induced to create a magnetic field that adjusts the net flux to a quantized value.

21.3   Josephson Junction

We can represent the order parameter in the polar form

$$\psi = \sqrt{n}e^{i\varphi} \tag{21.14}$$

As we have seen in Equation (20.22), a spatial gradient of φ generates a supercurrent, which physically is a current of Cooper pairs. By placing two different superconductors in contact, we have a *Josephson junction*. The difference in phases in the two superconductors constitutes a gradient that drives a supercurrent flowing across the junction.

Using Equation (21.14), we rewrite the current density [Equation (20.22)] as

$$\mathbf{j} = \frac{2e\hbar n}{m}\nabla\varphi - \frac{4e^2 n}{mc}\mathbf{A} \tag{21.15}$$

The first term is the supercurrent density. The order parameters of the two superconductors making up the Josephson junction, labeled 1 and 2, can be represented as

$$\psi_1 = \sqrt{n_1}e^{i\varphi_1}$$

$$\psi_2 = \sqrt{n_2}e^{i\varphi_2} \tag{21.16}$$

If $\varphi_1 > \varphi_2$, a current will flow from 1 to 2. In practice, a thin wafer of insulator is placed between 1 and 2, so that the Cooper pairs pass through the junction via quantum tunneling.

In the typical arrangement illustrated in Figure 21.2, the Josephson junction is a sandwich, with face area $0.025 \times 0.065 \text{ cm}^2$, and an insulating layer of thickness of

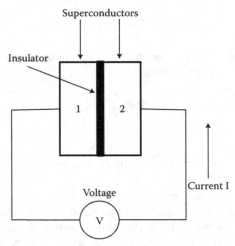

Figure 21.2   A Josephson junction.

2000 A. The entire assembly is maintained at a temperature of 1.5 K. A voltage V is applied between the superconductors, and a current I flows across the junction. The reference point for voltage is so chosen that the voltage is $-V/2$ at 1, and $V/2$ at 2. Since the tunneling link is very weak, we assume that the two superconductors are coupled linearly, and postulate the following equations for the order parameters, which are modeled after the NLSE neglecting the nonlinear self-interactions:

$$i\hbar \frac{d\psi_1}{dt} = eV\psi_1 + K\psi_2$$

$$i\hbar \frac{d\psi_2}{dt} = -eV\psi_2 + K\psi_1 \tag{21.17}$$

where K is a coupling constant. Bear in mind that the charge carriers are Cooper pairs with charge $-2e$. Our object is to study the current-voltage characteristic of the Josephson junction.

We now use the representation (Equation [21.16]). The first of the coupled equations becomes

$$\frac{i\hbar}{2\sqrt{n_1}} \frac{dn_1}{dt} - \hbar\sqrt{n_1}\dot{\varphi}_1 = eV\sqrt{n_1} + K\sqrt{n_2}e^{i(\varphi_2 - \varphi_1)} \tag{21.18}$$

where a dot denotes time derivative. Equating real and imaginary parts on both sides, and putting

$$\varphi = \varphi_2 - \varphi_1 \tag{21.19}$$

we obtain the equations

$$\frac{dn_1}{dt} = \frac{2K}{\hbar}\sqrt{n_1 n_2} \sin\varphi$$

$$\frac{d\varphi_1}{dt} = -\frac{eV}{\hbar} - \frac{K}{\hbar}\sqrt{\frac{n_2}{n_1}} \cos\varphi \tag{21.20}$$

A second set of equations can be obtained by interchanging the indices 1 and 2, and reversing the sign of V:

$$\frac{dn_2}{dt} = -\frac{2K}{\hbar}\sqrt{n_1 n_2} \sin\varphi$$

$$\frac{d\varphi_2}{dt} = \frac{eV}{\hbar} - \frac{K}{\hbar}\sqrt{\frac{n_1}{n_2}} \cos\varphi \tag{21.21}$$

From these we obtain

$$\frac{d\varphi}{dt} = \frac{2eV}{\hbar} - \frac{K}{\hbar}\left(\sqrt{\frac{n_1}{n_2}} - \sqrt{\frac{n_2}{n_1}}\right)\cos\varphi \tag{21.22}$$

Assuming $n_1 \approx n_2$, we neglect the second term on the right side and obtain

$$\frac{d\varphi}{dt} = \frac{2eV}{\hbar} \tag{21.23}$$

The Josephson current is defined by

$$I = e \frac{dn_1}{dt} = I_0 \sin \varphi \tag{21.24}$$

where

$$I_0 = \frac{2eK}{\hbar} \sqrt{n_1 n_2} \tag{21.25}$$

The current-voltage characteristic of the Josephson junction is therefore given by

$$I = I_0 \sin \varphi$$

$$V = \frac{\hbar}{2e} \frac{d\varphi}{dt} \tag{21.26}$$

As we shall see, the behavior of this system is most peculiar. If we apply a DC voltage, we get an AC current of such high frequency that it averages to zero. On the other hand, if we apply an AC voltage, we get a DC current.

21.4   DC Josephson Effect

Let us apply a constant voltage

$$V = V_0 \tag{21.27}$$

We can integrate $d\varphi/dt = 2eV_0/\hbar$ to obtain

$$\varphi(t) = \varphi_0 + \frac{2eV_0 t}{\hbar} \tag{21.28}$$

The current is given by

$$I(t) = I_0 \sin \left(\varphi_0 + \frac{2eV_0 t}{\hbar} \right) \tag{21.29}$$

For a typical voltage $V_0 = 10^{-3}$ v, we have

$$\frac{2eV_0}{\hbar} = 3.2 \times 10^{12} \ \text{sec}^{-1} \tag{21.30}$$

Therefore the current oscillates with an extremely high frequency and time-averages to zero:

$$I_{av} = 0 \tag{21.31}$$

21.5   AC Josephson Effect

Consider an AC voltage

$$V(t) = V_0 + V_1 \cos \omega t \tag{21.32}$$

We can write

$$\frac{d\varphi}{dt} = \omega_0 + \omega_1 \cos \omega t \tag{21.33}$$

where

$$\omega_0 = \frac{2e}{\hbar} V_0$$

$$\omega_1 = \frac{2e}{\hbar} V_1 \tag{21.34}$$

Thus

$$\varphi(t) = \varphi_0 + \omega_0 t + \frac{\omega_1}{\omega} \sin \omega t$$

$$I(t) = I_0 \sin \left(\varphi_0 + \omega_0 t + \frac{\omega_1}{\omega} \sin \omega t \right) \tag{21.35}$$

To simplify the notation, put

$$A = \phi_0 + \omega_0 t$$

$$B = \frac{\omega_1}{\omega} \sin \omega t \tag{21.36}$$

Suppose $\omega_1 \ll \omega_0$. Then

$$I(t) = I_0 \sin(A + B) = I_0(\sin A \cos B + \sin B \cos A)$$

$$\approx I_0(\sin A + B \cos A)$$

$$= I_0 \left[\sin(\phi_0 + \omega_0 t) + \frac{\omega_1}{\omega} \sin \omega t \cos(\phi_0 + \omega_0 t) \right]$$

$$= I_0 \sin(\phi_0 + \omega_0 t) + I_0 \frac{\omega_1}{2\omega} [\sin(\phi_0 + (\omega + \omega_0)t) + \sin(\phi_0 + (\omega - \omega_0)t)] \tag{21.37}$$

Upon time-averaging, all oscillating terms go to zero, and only the last term survives:

$$I_{av} = \begin{cases} 0 & (\omega \neq \omega_0) \\ I_0 \sin \varphi_0 & (\omega \approx \omega_0) \end{cases} \tag{21.38}$$

The average current as a function of frequency has a sharp peak at

$$\omega_0 = \frac{2eV_0}{\hbar} \tag{21.39}$$

Using the AC Josephson effect, the value of e/h can be measured to very high accuracy.

21.6 Time-Dependent Vector Potential

The presence of a time-dependent vector potential **A** gives rise to magnetic and electric fields given by

$$\mathbf{B} = \nabla \times \mathbf{A}$$

$$\mathbf{E} = -\nabla V - \frac{1}{c}\frac{\partial \mathbf{A}}{\partial t} \tag{21.40}$$

We can generalize the current-volatge characteristic [Equation (21.26)] by making the replacement

$$V \rightarrow V + \frac{1}{c}\frac{d}{dt}\int d\mathbf{r}\cdot\mathbf{A} \tag{21.41}$$

where the integral is taken along some path in space. Thus, the voltage equation becomes

$$\frac{d\varphi}{dt} = \frac{2e}{\hbar}V + \frac{2e}{\hbar c}\frac{d}{dt}\int d\mathbf{r}\cdot\mathbf{A} \tag{21.42}$$

Integrating this gives

$$\varphi(t) = \varphi_0 + \frac{2e}{\hbar}\int dt V + \frac{2e}{\hbar c}\int d\mathbf{r}\cdot\mathbf{A} \tag{21.43}$$

The current is then given by

$$I = I_0 \sin\left[\varphi_0 + \frac{2e}{\hbar}\int dt V + \frac{2e}{\hbar c}\int d\mathbf{r}\cdot\mathbf{A}\right] \tag{21.44}$$

All this is preparation for the next section.

21.7 The SQUID

The SQUID (superconducting quantum interference device) is a device that can measure magnetic flux, with enough sensitivity to detect one flux quantum. The arrangement is shown in Figure 21.3. Two Josephson junctions *a* and *b* are connected in parallel. The currents flowing through the junctions are respectively

$$I_a = I_0 \sin\left(\varphi_0 + \frac{2e}{\hbar c}\int_a d\mathbf{r}\cdot\mathbf{A}\right)$$

$$I_b = I_0 \sin\left(\varphi_0 + \frac{2e}{\hbar c}\int_b d\mathbf{r}\cdot\mathbf{A}\right) \tag{21.45}$$

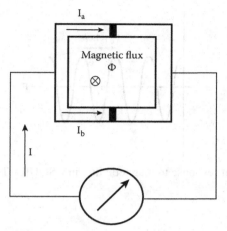

Figure 21.3 A SQUID consists of two Josephson junctions connected in parallel, forming a loop. Magnetic flux through the loop induces currents in the loop.

If there is a magnetic field $\nabla \times \mathbf{A}$ through the loop, then

$$\int_a d\mathbf{r} \cdot \mathbf{A} - \int_b d\mathbf{r} \cdot \mathbf{A} = \oint d\mathbf{s} \cdot \mathbf{A} = \Phi \qquad (21.46)$$

where Φ is the total magnetic flux going through the loop. We can put

$$\int_a d\mathbf{r} \cdot \mathbf{A} = \frac{1}{2}\Phi + C_0$$

$$\int_b d\mathbf{r} \cdot \mathbf{A} = -\frac{1}{2}\Phi + C_0 \qquad (21.47)$$

where C_0 is a constant that can be absorbed into φ_0. Thus

$$I_a = I_0 \sin\left(\varphi_0 + \frac{e\Phi}{\hbar c}\right)$$

$$I_b = I_0 \sin\left(\varphi_0 - \frac{e\Phi}{\hbar c}\right) \qquad (21.48)$$

Adding these, we obtain the total current

$$I = I_0 \left[\sin\left(\varphi_0 + \frac{e\Phi}{\hbar c}\right) + \sin\left(\varphi_0 - \frac{e\Phi}{\hbar c}\right)\right]$$

$$= 2I_0 \sin\varphi_0 \cos\left(\frac{e\Phi}{\hbar c}\right) = 2I_0 \sin\varphi_0 \cos\left(\frac{2\pi\Phi}{\Phi_0}\right) \qquad (21.49)$$

where Φ_0 is the elementary flux quantum given in Equation (21.13). As Φ changes, the current oscillates with a period Φ_0 as illustrated in Figure 21.4.

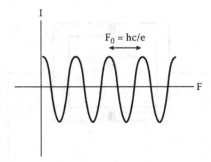

Figure 21.4 Current response to magnetic flux in a SQUID. The oscillation period is one flux quantum.

21.8   Broken Symmetry

We recall that the emergence of the superfluid order parameter is accompanied by a breaking of global gauge invariance. In the case of superconductivity, the symmetry under Equation (21.2) is a local instead of global gauge invariance, and it is spontaneously broken when the superconducting order parameter emerges. However, there is no Goldstone mode in this case. The nature of the electromagnetic coupling is such that the would-be Goldstone mode becomes the longitudinal component of the vector potential, which is needed to satisfy Equation (21.6). Now the photon acquires mass in the superconducting medium:

$$m_{\text{photon}} = \frac{\hbar}{c}\sqrt{\frac{16\pi e^2 n}{mc^2}} \tag{21.50}$$

which corresponds to the inverse of the penetration depth. Such a transfiguration of the Goldstone mode occurs whenever local gauge invariance is spontaneously broken. In the theory of elementary particles, this is called the "Higgs mechanism," and is supposed to be the mechanism that generates all the particle masses.

Problems

21.1 Consider a nonmagnetic superconducting medium filling the half space $x > 0$. Suppose at $x = 0^-$ there is a uniform magnetic field normal to the x axis. Find the magnetic field $B(x)$ in the medium.

21.2

(a) Consider a circuit in which a Josephson junction is connected in series with a resistor R, and a voltage source U_0. Set up the equation for the time development of φ, based on the current-voltage characteristic [Equation (21.26)].

(b) For $\kappa \equiv I_0 R / U_0 > 1$, show that the current approaches a limiting value I_∞, while $V \to 0$.

Reference

De Gennes, P.G., *Superconductivity of Metals and Alloys*, W.A. Benjamin, New York, 1966, Chapter 6.

Appendix

Mathematical Reference

A.1 Stirling's Approximation

In statistical analyses we frequently encounter the factorial

$$n! = 1.2 \ldots n \tag{A.1}$$

For large n, Stirling's approximation gives

$$n! \approx n^n e^{-n} \sqrt{2\pi n} \tag{A.2}$$

or

$$\ln n! \approx n \ln n - n + \ln \sqrt{2\pi n} \tag{A.3}$$

Usually the first two terms suffice. The relative error of this formula is about $(12n)^{-1}$, which is about 2% for $n = 4$.

To derive Stirling's approximation, start with the representation

$$n! = \Gamma(n+1) = \int_0^\infty dt\, t^n e^{-t} \tag{A.4}$$

The integrand has a maximum at $t = n$. The value at the maximum gives $n! \approx n^n e^{-n}$. Expanding the integrand about the maximum yields the corrections.

A.2 The Delta Function

The Dirac delta function $\delta(x)$ is zero if $x \neq 0$, and infinite if $x = 0$, such that

$$\int dx \delta(x) = 1 \tag{A.5}$$

where the range of integration includes $x = 0$. It is not a function in the ordinary sense, since its value is not defined where it is nonzero. We use it with the understanding that eventually it will find its way into an integral. Mathematicians call it a "distribution."

Some useful properties of the delta function are

$$\delta(x) = \delta(-x)$$

$$\delta(ax) = \frac{1}{a}\delta(x) \tag{A.6}$$

For any given function $f(x)$, we have

$$\int dx f(x)\delta(x-a) = f(a) \tag{A.7}$$

Its integral is the step function

$$\theta(x) = \begin{cases} 1 & (x > 0) \\ 0 & (x < 0) \end{cases} \tag{A.8}$$

The derivative of the delta function is the ϵ-function

$$\epsilon(x) = \begin{cases} 1 & (x > 0) \\ -1 & (x < 0) \end{cases} \tag{A.9}$$

The Fourier analysis is given by

$$\delta(x) = \int \frac{dk}{2\pi} e^{ikx} \tag{A.10}$$

In higher dimensions the delta function is defined as the product of 1D delta functions. For example,

$$\delta^3(\mathbf{r}) = \delta(x)\delta(y)\delta(z) = \int \frac{d^3k}{(2\pi)^3} e^{i\mathbf{k}\cdot\mathbf{x}} \tag{A.11}$$

A.3   Exact Differential

The differential

$$df = A(x, y)dx + B(x, y)dy \tag{A.12}$$

is said to be an "exact differential" if there exists a function $f(x, y)$ which changes according to the above when its independent variables are changed. We must have

$$A(x, y) = \frac{\partial f}{\partial x}$$

$$B(x, y) = \frac{\partial f}{\partial y} \tag{A.13}$$

Since differentiation is commutative, we have

$$\frac{\partial A}{\partial y} = \frac{\partial B}{\partial x} \tag{A.14}$$

This is the condition for an exact differential.

A.4 Partial Derivatives

Suppose three real variables x, y, z are constrained by one condition $f(x, y, z) = 0$, where f is a regular function. Partial derivatives have the properties

$$\left(\frac{\partial x}{\partial y}\right)_w \left(\frac{\partial y}{\partial z}\right)_w = \left(\frac{\partial x}{\partial z}\right)_w$$

$$\left(\frac{\partial x}{\partial y}\right)_w = \frac{1}{\left(\frac{\partial y}{\partial x}\right)_w} \qquad (A.15)$$

where w, the quantity being held fixed, is some function of the variables.

A.5 Chain Rule

The *chain rule* states

$$\left(\frac{\partial x}{\partial y}\right)_z \left(\frac{\partial y}{\partial z}\right)_x \left(\frac{\partial z}{\partial x}\right)_z = -1 \qquad (A.16)$$

This follows from

$$df = \frac{\partial f}{\partial x}dx + \frac{\partial f}{\partial y}dy + \frac{\partial f}{\partial z}dz = 0 \qquad (A.17)$$

where $\partial f/\partial x$ denotes the partial derivative with respect to x, with the other two variables kept fixed. From this we get

$$\left(\frac{\partial x}{\partial y}\right)_z = -\frac{\partial f/\partial y}{\partial f/\partial x}$$

$$\left(\frac{\partial y}{\partial z}\right)_x = -\frac{\partial f/\partial z}{\partial f/\partial y}$$

$$\left(\frac{\partial z}{\partial x}\right)_y = -\frac{\partial f/\partial x}{\partial f/\partial z} \qquad (A.18)$$

The desired relation is obtained by multiplying the above together.

A.6   Lagrange Multipliers

Consider a function of n variables $f(\mathbf{x})$, where $\mathbf{x}$ represents an n-component vector. An extremum of $f(\mathbf{x})$ is determined by

$$df(\mathbf{x}) = d\mathbf{x} \cdot \nabla f(\mathbf{x}) = 0 \qquad (A.19)$$

where $d\mathbf{x}$ represents an arbitrary infinitesimal change in x. If, however, there exists a constraint

$$g(\mathbf{x}) = 0 \qquad (A.20)$$

then dx cannot be completely arbitrary, but must be such as to maintain $dg = 0$, or

$$dg = d\mathbf{x} \cdot \nabla g = 0 \qquad (A.21)$$

This requires dx to be tangent to the surface $g = 0$, as represented schematically in Figure A1. We have to find the extremum of $f(\mathbf{x})$ on the surface $g(\mathbf{x}) = 0$.

An arbitrary differential $d\mathbf{x}$ can be resolved into transverse (orthogonal) and longitudinal (tangential) components with respect to the given surface:

$$d\mathbf{x} = d\mathbf{x}_T + d\mathbf{x}_L \qquad (A.22)$$

The condition for an extremum is

$$d\mathbf{x}_L \cdot \nabla f = 0 \qquad (A.23)$$

Since $d\mathbf{x}_L \cdot \nabla g = 0$ by definition, the above can be generalized to

$$d\mathbf{x}_L \cdot \nabla(f + \lambda g) = 0 \qquad (A.24)$$

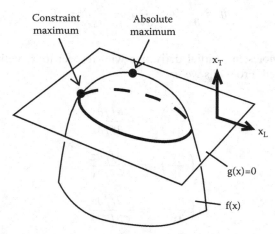

Figure A.1   Illustrating the method of Lagrange multipliers to find a constraint maximum.

where λ is an arbitrary number. Let us choose λ such that it satisfies

$$d\mathbf{x}_T \cdot \nabla(f + \lambda g) = 0 \tag{A.25}$$

Then, by adding this equation to the previous one, we have $d\mathbf{x} \cdot \nabla(f + \lambda g) = 0$, or

$$d[f(\mathbf{x}) + \lambda g(\mathbf{x})] = 0 \tag{A.26}$$

without any constraint. The parameter λ is called a Lagrange multiplier. To solve the problem, we can first find $\mathbf{x}$ for arbitrary λ. Denoting the result by $\mathbf{x}(\lambda)$, we determine λ by requiring $g(\mathbf{x}(\lambda)) = 0$. This determines the location of the desired extremum $\mathbf{x}(\lambda)$.

To summarize: the extremum of f subject to the constraint $g = 0$ can be found by finding the extremum of $f + \lambda g$, and then determine λ so as to satisfy the constraint.

A.7 Counting Quantum States

A free particle in quantum mechanics is described by the wave function

$$\phi_{\mathbf{k}}(\mathbf{r}) = \frac{1}{\sqrt{V}} e^{i\mathbf{k}\cdot\mathbf{r}} \tag{A.27}$$

where the wave vector $\mathbf{k}$ labels the state. This is normalized to one particle in the volume V. Periodic boundary conditions require

$$e^{i\mathbf{k}\cdot\mathbf{r}} = e^{i\mathbf{k}\cdot\mathbf{r}+i\mathbf{k}\cdot\mathbf{m}L} \tag{A.28}$$

where $\mathbf{m}$ is a vector whose components have possible values $0, \pm 1, \pm 2, \dots$. Thus $\mathbf{k}\cdot\mathbf{m}L = 0 \bmod(2\pi)$, and the spectrum of $\mathbf{k}$ is given by:

$$\mathbf{k} = \frac{2\pi\mathbf{n}}{L} \tag{A.29}$$

where $\mathbf{n}$ is a vector whose components have possible values $0, \pm 1, \pm 2, \dots$. Since the spacing between successive values is $2\pi/L$, the spectrum approaches a continuum as $L \to \infty$.

A sum over states is denoted by:

$$\sum_{\mathbf{k}} = \sum_{n_1=-\infty}^{\infty} \sum_{n_2=-\infty}^{\infty} \sum_{n_3=-\infty}^{\infty} \tag{A.30}$$

In the limit $L \to \infty$, this approaches a multiple integral. The spacing between values of k_1 is given by

$$\Delta k_1 = \frac{2\pi}{L} \Delta n_1 \tag{A.31}$$

and $\Delta n_1 = 1$. Thus

$$\sum_{n_1=-\infty}^{\infty} = \sum_{n_1=-\infty}^{\infty} \Delta n_1 = \frac{L}{2\pi} \sum_{k_1=-\infty}^{\infty} \Delta k_1 \rightarrow \frac{L}{2\pi} \int_{-\infty}^{\infty} dk_1 \qquad (A.32)$$

Therefore

$$\sum_{\mathbf{k}} \rightarrow \frac{V}{(2\pi)^3} \int d^3 k \qquad (A.33)$$

In terms of the momentum $\mathbf{p} = \hbar \mathbf{k}$, we have

$$\sum_{\mathbf{p}} \rightarrow \frac{V}{(2\pi\hbar)^3} \int d^3 p = \int \frac{d^3 r \, d^3 p}{h^3} \qquad (A.34)$$

This shows that the volume of a basic cell in phase space is

$$\tau_0 = h^3 \qquad (A.35)$$

A.8 Fermi Functions

The Fermi functions have the power series expansions

$$f_n(z) \equiv \sum_{\ell=1}^{\infty} (-1)^{\ell+1} \frac{z^\ell}{\ell^n} \qquad (A.36)$$

We illustrate the large z behavior by calculating that for $f_{3/2}(z)$. Go back to the integral representation (14.70), and put $y = x^2$, $z = e^\nu$:

$$f_{3/2}(z) = \frac{2}{\sqrt{\pi}} \int_0^{\infty} dy \frac{\sqrt{y}}{e^{y-\nu}+1} \qquad (A.37)$$

For ν, the factor $(e^{y-\nu}+1)^{-1}$, which is the occupation number, is nearly a step function, whose derivative is nearly a delta function. If we can rework the integrand into something involving that derivative, then most of the contribution to the integral would come from the neighborhood of the Fermi surface $y = \nu$. With this goal in mind, we make a partial integration:

$$f_{3/2}(z) = \frac{2}{\sqrt{\pi}} \left\{ \frac{2}{3} \left[\frac{y^{3/2}}{e^{y-\nu}+1} \right]_0^{\infty} - \frac{2}{3} \int_0^{\infty} dy \, y^{3/2} \frac{\partial}{\partial y} \frac{1}{e^{y-\nu}+1} \right\}$$

$$= \frac{4}{3\sqrt{\pi}} \int_0^{\infty} dy \, y^{3/2} \frac{e^{y-\nu}}{(e^{y-\nu}+1)^2} \qquad (A.38)$$

The integrand is now peaked at $y = v$. We put $y = v + t$ and obtain

$$f_{3/2}(z) = \frac{4v^{3/2}}{3\sqrt{\pi}} \int_{-v}^{\infty} dt \left(1 + \frac{t}{v}\right)^{3/2} \frac{e^t}{(e^t + 1)^2} \tag{A.39}$$

We are interested in large v, so the lower limit is near $-\infty$. Accordingly we write

$$f_{3/2}(z) = \frac{4v^{3/2}}{3\sqrt{\pi}} \int_{-\infty}^{\infty} dt \left(1 + \frac{t}{v}\right)^{3/2} \frac{e^t}{(e^t + 1)^2} + O(e^{-v}) \tag{A.40}$$

The asymptotic expansion is obtained by neglecting $O(e^{-v})$, and expanding the factor $(1 + \frac{t}{v})^{3/2}$ in inverse powers of $v = \ln z$:

$$f_{3/2}(z) \approx \frac{4v^{3/2}}{3\sqrt{\pi}} \int_{-\infty}^{\infty} dt \left(1 + \frac{3}{2}\frac{t}{v} + \frac{3}{8}\frac{t^2}{v^2} + \cdots\right) \frac{e^t}{(e^t + 1)^2} \tag{A.41}$$

In the power series, only terms of even power in t survive the integration. Thus

$$f_{3/2}(z) \approx \frac{4v^{3/2}}{3\sqrt{\pi}} \left(I_0 + \frac{3}{8v^2} I_2 + \cdots\right) \tag{A.42}$$

where

$$I_n = 2 \int_0^{\infty} dt \frac{t^n e^t}{(e^t + 1)^2}$$

$$I_0 = 1$$

$$I_2 = \frac{\pi^2}{3} \tag{A.43}$$

It is interesting to note

$$I_n = (n - 1)!(2n)(1 - 2^{1-n})\zeta(n) \quad (n \text{ even}) \tag{A.44}$$

where $\zeta(z) = \sum_{\ell=1}^{\infty} \ell^{-z}$ is the Riemann zeta function, a celebrated function of number theory.

For our purpose we only need the first few terms in the asymptotic expansion:

$$f_{3/2}(z) \approx \frac{4}{3\sqrt{\pi}} \left[(\ln z)^{3/2} + \frac{\pi^2}{8} \frac{1}{\sqrt{\ln z}} + \cdots\right] \tag{A.45}$$

Index

A

A and B coefficients, theory of, 239
Absolute temperature, 21–22
  Integrating factor of, 22–24
Absolute zero, 22
AC Josephson effect, 299
Adiabatic line
  Carnot cycle, within, 19
  Compressibility, 34
  Isotherm, *versus*, 18
Atomic nucleus, 11–12
Atoms
  Collisions, 66, 67, 70
  Diameter, 65
  Distribution functions, 70–71
  Distribution, most probable, 77–78
  Distribution, speed, 87
  Energy fluctuations, 93
  Flux, 99–100
  Identical particles, 197–198
  Kinetic energy, mean, 87
  Kinetic energy of, 66
  Macroscopic view, 65–67
  Occupation numbers, 198–200
  Random walk (*see* random walk)
  Spin state, 200
  Thermal wavelengths of, 195–196
Avogadro's number, 7, 65, 143

B

Big Bang, 12–13, 239
Black-body cavity, 237, 239. *See also*
  black-body radiation
Black-body radiation, 237, 239, 240, 246–247.
  See also black-body cavity
Boltzmann's constant, 74
Boltzmann's counting, correct, 74–75
Boltzmann's *H*, 76
Bose–Einstein condensation. *See also* Bose gas
  Clapeyron equation, 260
  Condensate, 253–254
  Description, 251
  Equation of state, 254–255

Ideal Bose gas, of, 281
Liquid helium, in, 259–260
Macroscopic occupation, 251–253
Observations of, 285, 286*f*
Phase diagram, 252*f*
Phase formation, 257–258
Specific heat of, 256
Bose enhancement, 239–240, 246.
  See also Bose gas; Bose–Einstein
  condensation
Bose gas. *See also* Bose–Einstein
  condensation; bosons
  Debye specific heat, 243
  Electronic specific heat, 244–246
  Enhancement, 239–240, 246
  Ideal, 255
  Maxwell–Boltzmann distribution, 239–240
  Phonons, 241–242, 245
  Photons density, 237–239
  Planck distribution, 238
  Spontaneous emission, 240
  Stimulated emission, 240
Bose statistics, 197, 203, 221. *See also* Bose
  gas; bosons
Bosons, 197. *See also* fermions
  Classical limit, 208–209
  Entropy, 206
  Equation of state, 207–208
  Occupation numbers, 198, 219, 220
  Parameters, 204–205
  Photon bunching, 220–221
British thermal unit (Btu), definition of, 6
Brownian motion, 136–138, 141*f*
  Conservation of particles, 142
  Einstein, importance to, 138–139
  Forced oscillator in medium example,
    189–192
  Gaussian distribution, *versus*, 145–146
  Langevin equation (*see* Langevin equation)
  Stock market, model of, 145–146, 147*f*

C

Cahn–Hilliard equation, 60, 278
Campbell's theorem, 161–162

313

Canonical ensemble, 113, 114*f*
  Classical, 111–113, 118–119
  Energy fluctuations, 115–116
  Grand (*see* grand canonical ensemble)
  Microcanonical (*see* microcanonical
    ensemble)
  Quantum partition function, 216–217
  Thermodynamics, relationship between, 115
Carnot cycle, 19–20
  TS diagram, 38
Carnot's principle, 143
Central limit theorem, 140
Chain rule, 307
Champman-Kolmogorov equation, 171
Clapeyron equations, 50–51, 60
Classical canonical ensemble, 111–113
  Ideal gas, of, 118–119
Clausius, 20, 22
Clausius's theorem, 22–24, 39
Coefficient of thermal conductivity, 105–106
Collisionless regimes, 99–100
Conservation laws, 101
Correct Boltzmann's counting, 74–75
Critical opalescence, 128
Curie's law, 9

D

DC Josephson effect, 298
Debye frequency, 242
Debye function, 243
Debye model, 241–242, 243, 247
Debye specific heat, 243
Debye temperature, 242
Degeneracy, 92
Delta function, 305–306
Demon, Maxwell's, 101
Diffusion, 140
Diffusion constant, 69, 103–105
Dirac delta function, 305–306
Dissipation, definition of, 144
Distribution entropy, 76, 78, 88–89
Distribution function of atomic systems, 70–71
Dulong and Petit, law of, 243

E

Effusion, 99, 101
Einstein's relation, 142, 187–188
Einstein's theory of 1905, 138–139, 140, 152
Einstein, Albert, 138–139
Electron donor levels, 233
Energy
  Exchange of, 111, 112
  Fixed, 111

Free, identifying, 207
Free, minimizing, 116–118
Energy surface, 70
Entropy
  Calculating, 206
  Distribution (*see* distribution entropy)
  Ideal gas, of, 26–27, 81–82
  Irreversible isothermal expansion,
    relationship between, 35–37
  Isolated system, of, 25
  Loss, relationship between, 35–37
  Shannon theory (*see* Shannon entropy)
Equation of state, 207–208
Equipartition of energy, principle of, 85–87
Ergodic hypothesis, 72
Exact differential, 306

F

False vacuums, 12–13
Fermi energy, 225–226
Fermi functions, 310–311
Fermi gas. *See also* fermions
  Ground state, 226–227
  Holes, 230–231
  Low temperature properties, 228–230
  Particles, 230–231
  Thermal equilibrium of, 230–231
  Zero-point pressure, 227
Fermi statistics, 197
Fermi surface, 230–231
Fermi temperature, 227–228
Fermi wave number, 225
Fermions, 197. *See also* bosons; Fermi gas
  Classical limit, 208–209
  Entropy, 206
  Equation of state, 207–208
  Fermi statistics, 202–203
  Free, 225–226
  Number of, in quantum state, 230–231
  Occupation numbers, 198, 199, 219
  Parameters, 204–205
  Slater determinant, 199
Ferromagnetic systems, 9–10
Feynman path integral, 181
First-order phase transitions
  Gas-liquid, 48*f*
  Isotherm, of an, 48*f*
  Liquid-solid, 48*f*
  Overview, 47
  Phase coexistence, 49–50
  Triple point, 49
Fluctuation, definition of, 144
Fluctuation-dissipation theorem, 142,
    144–145, 187

Fokker–Planck equation, 148–149, 172–172, 184, 188–189
Fourier analysis
  Power spectrum, 166, 168
  Time series, of, 165–166

G

Gas
  Liquid transition, 56–57
  Thermal equilibrium, 65
  Uniform velocity movement, 87
  Velocity, 106
Gaussian distribution
  Brownian motion, *versus,* 145–146
  Poisson distribution, relationship between, 155–156
Gibbs potential, 41
Ginsburg–Landau theory, 266–267, 278, 293
  Microscopic theory, relationship between, 267–268
Goldstone mode, 283, 289–290
Grand canonical ensemble
  Density fluctuations, 127–128
  Ensemble average, 123–124
  Grand partition function, 123–124, 217–219
  Pair creation, 128–130
  Parametric equation of state, 126–127
  Particle average, 124–125
  Particle reservoir, 123
  Thermodynamics, relationship between, 125–126
  Viral expansion, 127

H

Hamiltonian equations of motion, 69–70, 73
Heat
  Latent, 47, 49
Heat bath method, 8, 177
Heat capacity 5–7
Heat conduction, 105–106
Heat reservoir, 8, 177
Heat transfer, 5–7, 22t
Helmholtz free energy, 40–41
Hydrodynamics, 99, 100
  Nonviscous, 101–102
  Sound wave equations, 103

I

Ideal gas
  Energy equation, 34
  Entropy of, 26–27, 81–82
  Law, 7
  Occupation number fluctuations, 219
  Pressure of, 84–85, 205–206
  Properties of, 16–18
  Temperature, 7
  Thermodynamics, 89–90
Inflationary universe, 12
Information theory, 78–80, 101
Ising model, 176–178
Isothermal compressibility, 34
Isothermal expansion, 36
Isotherms, 8

J

Josephson junction, 296–298, 300–301
Joule, James Prescott, 16
Joules, definition of, 6

K

Kinetic theory, 2

L

Lagrange multipliers, 77, 78, 308–309
Landau free energy, 274
Langevin equation
  Diffusion coefficient, 187–188
  Einstein's relation, 188
  Energy balance of, 185–186
  Fluctuation-dissipation theorem of, 187
  Fourier transforms, solved using, 184
  Oscillator/liquid example, 189–192
  Overview, 183–185
  Transition probability, 188–189
Latent heat, 47, 49
Legendre transformation, 41
Liquid helium, 259–260
Liquid, heating of, through stirring, 189–192
Lorentzian distribution, 168

M

Macroscopic bodies
  Exceptions of intensive/extensive classifications, 2
  Extensive quantities, 2
  Intensive quantities, 2
  Thermal equilibrium of, 2
Magnetic flux quantum, 294–295
Magnetic systems
  Overview, 9–11
Markov process, 171, 174

Matter, atomic structure of, 1
Maxwell relations, 39, 40, 41, 42, 42*f,* 43
Maxwell–Boltzmann distribution, 78,
    231
  Distribution entropy, 88–89
  Equipartition of energy, relationship
    between, 85–87
  Fluctuations, 90–91
  Ideal gas pressure, 84–85
  Parameters of, 83–84, 85
  Probable distribution, 90–91
  Speed distribution, 87, 88*f*
Maxwell's demon, 101
Mean free path, 65
Meissner effect, 293–294
Metropolis algorithm, 175
Microcanonical ensemble, 73–74
  Overview, 111, 114*f*
  Quantum states of, 201–202
Molar volume, 65
Molecular reality, 143–144
Monte Carlo method, 173–175,
    176–178
Motion equations, 92

N

N-type semiconductor, 233
Navier–Stokes equation, 107–108
Noise
  Definition, 168
  Thermal (*see* thermal noise)
Noise, shot, 157–160, 179
Non-Carnot cycle, 38
Nyquist noise, 134–135, 136*f*
Nyquist theorem, 135, 148

O

Occupation number, 71
Order parameter
  Cahn–Hilliard equation
    (*see* Cahn–Hilliard equation)
  Correlation length, 274–276
  Critical exponents, 273–274
  First-order phrase transition, 277
  Functional differentiation, 268–270
  Functional integration, 268–270
  Ginsburg–Landau theory, 266–267
  Mean field theory, 271–273, 274
  Microscopic theory, 267–268
  Overview, 265
  Phase transitions, essence of, 265
  Second-order phase transition,
    270–271

P

P-type semiconductor, 233
Paramagnets, 46
Partial derivatives, 307
Perrin, M.J.
  Brownian motion, 137*f,* 140, 141*f*
  Molecular reality, 143–144
  Nobel prize, 136
  Noise of atoms, 136–137
Phase space, 69–70
Phase transitions
  First order (*see* first-order phase transitions)
  Overview, 265
  Second order (*see* second-order phase
    transitions)
Phonons, 241–242, 245
  Goldstone mode, 289–290
Poincare, Henri, 145
Poisson distribution, 154–155
  Gaussian distribution, relationship between,
    155–156
Probability
  Basic properties of, 151–152
  Binomial theorem, 153

Q

Quantum degeneracy temperatures, 196
Quantum mechanics
  Classical mechanics, *versus,* 213
  Density matrix, 214–215
  Free particle counting, 309–310
  Incoherent superposition of states,
    213–214
Quarks, 1

R

Random walk, 67–69, 72, 104
Randomness, 151
  Shot noise, of, 157–160, 179
Representative point, 69
Reynolds number, 108

S

Sacker–Tetrode equation, 43
Second-order phase transitions, 62–63,
    270–271
  Description, 47
Semiconductors, 233–235
Shannon entropy, 78–80
Shot noise. *See* noise, shot
Single-particle quantum numbers, 198

Slater determinant, 199
Smoluchowski equation, 171
Solids
  Einstein model of, 247
  Electrons in, 231–233
  Lattice of, 244
Sound waves, 103
SQUID. *See* superconducting quantum
      interference device (SQUID)
Statistical ensemble, 72–73
Statistical mechanics, definition of, 2
Stefan's constant, 238
Stefan's law, 238, 246–247
Stirling approximation, 118, 202, 305
Stochastic processes
  Binomial distribution of variables,
      152–153
  Central limit theorem, 157
  Definition as related to physics, 151
  Gaussian distribution, 155–156
  Markov process (*see* Markov process)
  Overview, 151–152
  Poisson distribution of variables, 154–155
  Shot noise, randomness of, 157–160
  Time-series (*see* time-series analysis)
Stoichiometric coefficients, 131–132
Superconducting quantum interference device
      (SQUID), 300–301
Superconductivity
  AC Josephson effect, 299
  Broken symmetry, 302
  DC Josephson effect, 298
  Josephson junction, 296–298, 300–301
  Magnetic flux quantum, 294–295
  Meissner effect, 293–294
  Superconducting quantum interference
      device (SQUID), 300–301
  Time-dependent vector potential, 300
Superfluidity
  Condensate wave function, 281–282
  Flow, 287–289
  Global gauge invariance, 283
  Mean-field theory, 284–285
  Quantum phase coherence, 286–287
  Spontaneous symmetry breaking, 282–284
  Velocity, 288
  Vortex, 288

T

Theory of A and B coefficients, 239
Thermal ensemble, 173–175
Thermal equilibrium, 2, 133
  Atomic collisions, 65–67, 99
  Equation of state, 4

Fermi gas, of (*see* Fermi gas)
Liquid-vapor, 61
Thermal expansion, 34
Thermal noise
  Diffusion of, 140
  Importance of, 133
  Nyquist noise, 134–135, 136*f*
Thermal wavelengths, 195–196
Thermodynamic limits, 27
  Free particles, of, 199–200
  Intensive quantities used, 4
  Variables of, 3–4
Thermodynamic transformation, 4–7,
      42–43
  Internal energy, of, 4–78–9
Thermodynamics. *See also* thermal
      equilibrium; thermodynamic limits;
      thermodynamic transformation
  Classical, 136
  Definition, 2
  Energy equation, 33–34
  First law, 8–9, 16–18
  Heat equations, 15–16
  Ideal gas, of (*see* ideal gas)
  Internal energy, 114
  Quantities, 27
  Second law, 19, 20–21, 30–31, 92,
      136, 143
Thomas–Fermi approximation, 285
Time's arrow, 92–93
Time-reversal invariance, 92
Time-series analysis
  Decomposition of, into sinusoidal
      components, 165–167
  Detailed balance, 170
  Ensemble average, 164–165
  Fokker–Planck equation (*see* Fokker–Planck
      equation)
  Fourier analysis, 165–166
  Markov process, 171, 174
  Noise, relationship between, 168–169
  Overview, 163–164
  Power spectrum, 166, 168
  Signal, relationship between, 168–169
  Transition probabilities, 170

V

Vacuums, false, 12–13
Valence band, 231, 233
Van der Waals equation of state, 51–53
  Critical point, 53–54
  Density, assumptions regarding, 57–58
  Maxwell construction, 55–56, 57
  Nucleation, 58, 59*f*

Scaling, 56–57
Spinoidal decomposition, 58, 60
Virial expansion, 53
Vapor pressure, 47
Virial expansion, 53
Viscosity, 106–107, 108
Volatility, 145

W

Wiener–Kintchine theorem, 166, 167, 168*f*

Z

Zero fields, 10

Physical constants

Planck's constant	$h = 6.626 \times 10^{-27}$ erg s
	$\hbar = h/2\pi = 1.054 \times 10^{-27}$ erg s
Velocity of light	$c = 3 \times 10^{10}$ cm s^{-1}
Proton charge	$e = 4.803 \times 10^{-10}$ esu $= 1.602 \times 10^{-19}$ C
Proton mass	$M_p = 1.673 \times 10^{-24}$ g
Electron mass	$m_e = 9.110 \times 10^{-28}$ g
Avogadro's number	$A_0 = 6.022 \times 10^{23}$
Boltzmann's constant	$k_B = 1.381 \times 10^{-16}$ erg K^{-1}
Gas constant	$R = 8.314 \times 10^{-7}$ erg K^{-1}mole^{-1}

Conversion

Length, weight, temperature

1 litre $= 10^3$ cm^3

1 in $= 2.54$ cm

1 lb $= 0.453$ kg

$X°C = (273.15 + X)$ K

$X°F = \frac{5}{9}(X - 32)°C$

Energy

1 J $= 10^7$ erg

1 eV $= 1.602 \times 10^{-12}$ erg

1 eV$/k_B = 1.160 \times 10^4$ K

1 cal $= 4.184$ J

1 Btu $= 1055$ J

1 ft lb $= 1.356$ J

1 kwh $= 3.6 \times 10^6$ J

Power

1 W $= 1$ J s^{-1}

1 hp $= 550$ ft lb s$^{-1} = 746$ W

Pressure

1 atm $= 1.013 \times 10^6$ dyn cm$^{-2} = 760$ mm Hg

1 bar $= 750$ mm Hg $= 10^6$ dyn cm^{-2}

1 torr $= 1$ mm Hg $= 1333$ dyn cm^{-2}

1 Pa $= 10^{-5}$ bar $= 10$ dyn cm^{-2}

Water at STP

Density $= 1$ g cm^{-3}

$C_P = 1$ cal g^{-1} K^{-1}